Contents

What's Next?

www.The60MinuteStartup.com

Start your business today the agile way! As a valued reader, you get all the templates, scripts, and tools you need to build your 60-minute startup, all at no cost to you.

Just visit this special web page for your free content upgrades so you can set up your business, build a website and other online assets, attract paying customers, and more in the shortest time possible.

Want support from other entrepreneurs on this thirty-day journey? Join the free private group to ask questions, bounce around ideas, and possibly even find your first paying customers. Go to the link below.

www.The60MinuteStartup.com

Tell Me What You Think

Let other readers know what you thought of *The 60-Minute Tech Startup*. Please write an honest review for this book on Amazon or on your favorite online bookshop.

Twelve of the entrepreneurs featured in this book appear on The Agile Entrepreneur podcast with author Ramesh Dontha. Listen to at

www.RameshDontha.Com/podcast

The Secret to Launch a Successful Side Hustle

Alex and Sasha both work in tech. And they both want to start a business.

Alex thinks he has to quit his stable full-time job that pays well if he's ever going to have a shot at building the next Facebook, Salesforce, or Uber. Sasha believes it's possible to have the best of both worlds, enjoying the luxury of secure employment while translating her existing skill set into a fun side hustle.

Alex believes that a tech startup requires both venture capital (VC) and sweat equity. If his business idea is going to amount to anything sellable in five years, he's going to have to hustle, grind, and eat ramen for breakfast. Sasha knows that many billion-dollar Silicon Valley superstars like Twitter, Groupon, and Oculus started as bootstrapped side hustles with little to no outside investment.[1]

Alex gets up at 4:30 in the morning and works through his commute, drafting a comprehensive go-to-market plan, perfecting his investor pitch deck, and meticulously researching venture capitalists on AngelList, Crunchbase, and LinkedIn. Meanwhile, Sasha puts one hour a day into her business after she tucks her daughter into bed. She has her proof of concept built, project and contract templates drafted, and her first paying users signed up by the end of her first month in business.

Alex quits his six-figure eight-to-five job after scheduling his first sit-down with a VC. He moves with his girlfriend into a studio apartment. He promises her a Malibu mansion once that startup seed money comes through . . . which it most definitely will . . . any day now. Well into month two in business, Sasha has already made her first thousand dollars. To scale her business, she offers clients a done-for-you retainer service and creates an affiliate program to incentivize referrals. After paying off old credit card debt, she uses her now-reliable second income to pay off her student loan early.

---

[1] Haden, Jeff. "21 Side Projects That Became Million-Dollar Startups (and How Yours Can Too)." Inc., November 16, 2017. www.inc.com/jeff-haden/21-side-projects-that-became-million-dollar-startups-and-how-yours-can-too.html.

Alex's first VC presentation doesn't go as hoped. Where's his prototype? His beta users? His hockey stick projections? Yet Alex leaves feeling motivated to work harder. He drains his savings to build a prototype offer. It's all going to be worth it one day, he reassures his girlfriend, who picks up a second job at the corner coffee shop. But when the working prototype is finished, the VC doesn't return Alex's calls. So he buys courses and attends workshops on creating a better slide deck. Across town, Sasha has officially doubled her income. With her side hustle hauling in the same as her day job, she gives herself a target date to quit and go full time in her proven venture. But first, she's going to save up for a down-payment on her first home.

The path Alex took to start a tech company is the one most aspiring entrepreneurs assume they have to follow. It's painstakingly slow, requires a huge (or total) pay cut, and makes no guarantees. It's high risk with the hope of high reward, but more often than not, the reward never comes. Sooner or later, founders like Alex have to choose between chasing a dream and paying their bills. But side hustlers like Sasha go a different way, the path of lowest risk with the highest upside. The only objectives and key results (OKRs) she cares about in the beginning revolve around getting customers. So while Alex's tech business drags him behind in life, Sasha's catapults her so much further ahead that Alex will probably never catch up.

Maybe Alex will miraculously find himself in the right place at the right time talking to the right person who just so happens to be an investor looking for a new business to fund.

Or maybe not.

What Agile Software Development Can Teach Us about Successful Startups

If you work in the technology industry, you've no doubt heard of agile software development. It's a project management methodology that emphasizes speed, flexibility, and results over perfection, precision, and tradition. But you may not be familiar with the origin story of the agile method—or how agile translates into the fastest way to start a tech company and get paying customers without pursuing outside funding or quitting your day job.

In February 2001, seventeen software developers from around the world met for a three-day retreat to get drunk over problems with their bosses and share ideas to get projects done on time.[2] This was a casual event. Little planning went into it beyond booking the lodge. Yet what emerged from their conversations was the Agile Software Development Manifesto. This profound document changed the way developers plan projects, collaborate with coworkers, report to management, and meet deadlines. The Manifesto states four primary values that should guide every project:

---

[2] History: The Agile Manifesto. Accessed April 7, 2020. www.agilemanifesto.org/history.

1. Individuals and interactions over processes and tools

2. Working software over comprehensive documentation

3. Customer collaboration over contract negotiation

4. Responding to change over following a plan[3]

The agile approach revolutionized software, made Silicon Valley the tech capital of the world, and made possible the Airbnbs, Dropboxes, and Instagrams we can't imagine living without. If agile software founders can build companies worth billions in a garage, imagine what agile entrepreneurship can do for you! Well, you don't have to imagine—*you* can be the next Sasha. Agile makes it possible.

Let's back up a second and see how Merriam-Webster defines the word "agile."

1. marked by the ready ability to move with quick and easy grace.

2. having a quick, resourceful, and adaptable character.[4]

Typical high-tech startup founders invest years and fortunes building their business, which may never produce a profit. Agile entrepreneurs are different. They believe that *good enough to make money is good enough to make money.* Agile entrepreneurs choose real-life feedback over business

---

[3] Eby, Kate. "Comprehensive Guide to the Agile Manifesto." Smartsheet, July 29, 2016. www.smartsheet.com/comprehensive-guide-values-principles-agile-manifesto.

[4] "Agile." Merriam-Webster. Merriam-Webster. Accessed April 7, 2020. www.merriam-webster.com/dictionary/agile.

school theories. Fast over methodical. Done now over done perfectly. Money in the bank over money from VCs.

The agile method is so superior to every other path to starting a business and getting your first paying customers that I dedicated my first book to it, *The 60-Minute Startup: A Proven System to Start Your Business in 1 Hour a Day and Get Your First Paying Customers in 30 Days (or Less)*. The book teaches how to inventory your interests, passions, and business ideas and turn them into a viable business that attracts paying customers by the end of month one.

The book you're reading now—*The 60-Minute Tech Startup*—marks the first in a series of spinoffs that address the unique challenges and also opportunities of specific industries. What it takes to build a high-tech side business without upsetting your employer looks very different from quitting your job to freelance full time in a totally different industry.

If you're not looking to start a technology company, I highly recommend that you also read the first book in this series, *The 60-Minute Startup*. So we're clear on what is in scope for the book you're reading now, let me define what I mean by tech company. I like Wikipedia's definition.

> A technology company (often tech company) is a type of business entity that focuses mainly on the development and manufacturing of technology products or providing technology as a service.[5]

---

[5] "Technology Company." Wikipedia. Wikimedia Foundation, April 6, 2020. www.en.wikipedia.org/wiki/Technology_company.

I'm passionate about helping you turn your tech skills into a viable business because I've done that for myself. And I can tell you that working for yourself is the most satisfying way to earn a living irrespective of how rewarding your day job may be.

I got trained as a Mechanical Engineer and an Industrial Engineer (but never worked for even a single day as either). I started my career as a Systems Analyst and traveled in India, Europe, and the USA. After a while, I realized that my interests were elsewhere, so I left that cushy job and got my MBA. I then worked in marketing, management, business development, and strategic planning areas for Fortune 100 companies, enjoyed the work, learned (and earned) a lot, and traveled the world all over again.

Throughout all these years of paying my dues to the corporate world, I dreamed of being an entrepreneur. I started my first business, a tutoring business, just after my undergrad college while waiting to figure out what I wanted to do with my life. That venture didn't last long. Since then, I started four different companies, went through the ups and downs of being an entrepreneur, sold two of them, and continued to run my remaining ventures. Today, I currently actively run a strategic management consulting business focusing on data analytics. You can get more information on that at www.DigitalTransformationPro.com.

Based on my experience and that of guests on my popular startup podcast The Agile Entrepreneur, I figure that you probably have a lot of questions about tech entrepreneurship, questions such as:

- Where do I start?
- Can I start a business while I am still working in my day job?
- How long before I can get my first paying customers?
- How long before I can quit my day job?

You'll find the answers to these and other questions in the pages ahead. For now, let's tackle the first concern, where to start.

As you know by this point, I'm a huge fan of starting your own business while still working in your full-time job. This solves two problems:

1. Your financial risk will significantly decrease as you'll continue to have salary and benefits, and
2. your stress levels will go down significantly as well.

Unfortunately, many people who start a tech business on the side end up like Alex. They take too much time to launch and build their side business, as time at their day job can be used as an excuse to move slowly. My recommendation is to have a strict timeline to start and launch your side business so you do not get stuck in the perennial loop of learning.

In *The 60-Minute Startup*, I introduced the fastest way to start your business based on the agile method. Set aside sixty minutes a day for thirty days to work on your business, and you should be able to get your first paying customers within that time.

Because you work in tech and will continue to do so in your side hustle, your timeline will look a little different. You'll use the thirty days to test your new product or service, get feedback, validate the business idea, and start generating income before you think about quitting your job. Again, you'll apply agile principles to do this with less stress, less money, and in less time.

Is a Side Hustle a Side Hustle Forever?

Will *The 60-Minute Startup* confine you to the life of a solopreneur where you're a one-person show whose revenue peaks at six figures? No. Aspiring entrepreneurs like Alex think that a tech business should start as a million-dollar business right off the bat and that a side hustle will forever remain a side hustle—so pick one. This is a false dichotomy. I'd like to provide you with confidence that your side hustle *can* become the unicorn you dream of.

Remember that companies like Twitter and Groupon started as their founders' side gigs. If you've ever written a line of code, you know about GitHub, the world's largest community of developers to discover, share, and build better software. As the story goes, co-founders Chris Wanstrath and PJ Hyett were working as website developers when they figured there had to be a better way to work with open source code. So they worked nights and weekends to build their own platform. By the end of its first year online, GitHub had accumulated over 46,000 public repositories. Within two-and-a-half years, the company had 100,000 users. After ten years in business, Chris and PJ sold Github to Microsoft for $7.5 billion. Wow!

While GitHub's co-founders did raise money to help grow their business, outside funding is not a requirement to build a massively successful tech company. Chances are you've heard of Udemy, the elearning platform for students and adult professionals. Perhaps you've even taken a Udemy course or two to bolster your own tech skills. Udemy co-founder Gagan Biyani worked on the company as a side hustle because he eventually wanted to leave his stressful job at consulting firm Accenture. Gagan and two colleagues who joined him on the project tried to raise venture capital funding, but they were rejected thirty times. So they decided to bootstrap the business, leveraging their own skills and savings to build their proof of concept. A few months into launch, Udemy had amassed 1,000 instructors, 2,000 courses, and nearly 10,000 registered users. With the business idea validated, Gagan and his co-founders then sought out VC money to continue growing. Their first outside investment was $1 million, but they've since raised over $170 million to date.

How do you go from a profitable side hustle to a venture capital-backed venture? That question is outside the purview of this book. Because we don't want to get ahead of ourselves. According to Crunchbase, the number one reason businesses fail is no market need.[6] Potential customers don't want what you're selling. Having invested in and even bought businesses before, I can tell you what investors care about—customers. If your business is not able to attract customers, it's not a business worth investing in. So long before we think about attracting investors for your tech company, we need to build a viable business, and that means getting customers fast. Don't put the cart before the horse, the investors before the customers.

Maybe you don't want to worry about pleasing venture capitalists, and you'd like to bootstrap this thing forever. In that case, *The 60-Minute Tech Startup* is perfectly suited for you. Because you'll be getting your first paying customers, clients, or users within thirty days, you'll soon be able to self-fund your project. You'll be in control of your own growth. So if you'd like to one day work ten extra hours a week earning $10,000 a month while you keep your day job, great! You can do that. In later chapters, you'll learn how to go from working with two or three customers to dozens or more, if you want to. That's how you'll transition from earning a few hundred dollars extra to thousands.

Which brings me to how this book works. Before you start any business, it first makes sense to plan out what steps you'll take, when, and in what order. So let's talk about that next.

---

[6] "Why Startups Fail: Top 20 Reasons." CB Insights Research, February 3, 2020. /www.cbinsights.com/research/startup-failure-reasons-top/.

How to Use *The 60-Minute Tech Startup*

60 minutes a day X 30 days = 1 viable business

That's my promise to you if you read and apply everything in the pages to come. As I wrote this book, I skipped all the typical (bad) startup advice and trimmed essential tasks down to the *most* essential. Ever heard of the 80/20 rule? 80 percent of the results you want come from 20 percent of your effort. Well, *The 60-Minute Tech Startup* like its predecessor *The 60-Minute Startup* is more like a 99/1 rule. That means you're doing one thing here and one thing there that has a big impact. I can tell you from experience that what *feels* productive often isn't. The only important activities to start and build a business aren't the ones that take days and weeks to finish. In fact, they don't take as long as you might expect. Only a few tasks lead to paying customers. For aspiring tech entrepreneurs like you, that is the only goal, customers. No customers, no business.

This book is meant to be read one chapter a day for thirty consecutive days. In each chapter, I'll show you what to do next and how to get it done fast. Each chapter opens with a story of a successful tech entrepreneur who took the agile approach to that day's activities. That way you can see for yourself what that task looks like when it's done right. You'll also see a checklist of steps for the day, suggested time for each, and any templates you need to get the day's work done in sixty minutes or fewer. If you read one chapter every day and complete the tasks, you will most likely have paying customers in thirty days. For any tech tasks that change over time (e.g., using a project management platform), I'll direct you to The60MinuteStartup.com for my most up-to-date resources. That way you can start your own business today, whether you're reading this book in 2020, 2030, or beyond!

Over the next month together, we're going to borrow principles from the Agile Software Development Manifesto and apply them to your tech startup dream. We'll do so using something called a *scrum*. Our scrum is the recipe we'll follow that tells us what essential ingredients we need for our agile framework. Agile is the *what*, *scrum* is the how. Let's break down the critical elements of our scrum:

- Sprint
- Sprint team
- Product backlog
- Sprint backlog
- Story points
- Sprint planning
- Daily standup
- Sprint review
- Sprint retro

In a scrum, you **sprint** to get meaningful work accomplished over a defined time period. In *The 60 Minute Startup*, we'll have two fifteen-day sprints so we'll finish in thirty days exactly.

A **sprint team** is composed of the owner, the scrum master (a servant-leader who manages and coaches the team while tracking deliverables), and anyone else working on the scrum. In our case, the owner is you, the scrum master is me, and the rest of the team is anyone who helps you accomplish the day's task or the week's sprint (a lawyer, a copywriter, etc.). You're taking a team approach in everything you do. You're not in this alone. In every chapter, you're going to witness how another agile entrepreneur accomplished the same tasks you're going to do that day. You're essentially joining a sprint team of superstars. That way you'll start and finish each day feeling motivated to keep going.

The **product backlog** is simply a list of the requirements needed to develop a meaningful product. In our case, the product backlog is the entire list of all the tasks you'll accomplish in this book over the next thirty days to start your business and get paying customers. You're prioritizing the important things. Your only objective right now is to get your first paying customers. Prioritize every task that gets you closer to that goal. For example, since you're keeping your day job, is it a good idea to build a website? At my data analytics company, most clients find me through an internet search. So I've prioritized a website. For your side business, a LinkedIn or other social media profile page might do the job, so a website should wait, especially if it gets you into hot water with your employer. We'll address this exact issue on day one.

When you start a business, figuring out what to do and in what order is one of the hardest tasks. That's why I've defined what the highest priority actions are and laid them out for you in the order you should do them.

The **sprint backlog** is the list of things that need to be accomplished during each sprint. In our case, the first fifteen-day sprint covers your product or service, your value proposition, your proof of concept, your physical and digital workspace, and everything else you need to start working with customers, clients, or users, including contract templates, a pricing guide, and the right project management system.

The second fifteen-day sprint is all about attracting and engaging with potential buyers, handling the selling and negotiation conversations professionally, and leveraging those relationships to generate new business so you can build a profitable, sustainable startup. Congratulations! You now have a real tech company.

Story points are essentially a breakdown of each task into steps with how long it takes to do each step. Don't worry about being exact. An estimate is good enough. Knowing what you know now, what kind of turnaround can you expect on this step? You make an initial estimate of how much effort you'll need to put into a task, then you keep updating that estimate based on how things are going. Every day I estimate that the tasks I give you in each chapter will take sixty minutes. If one day's tasks go quicker, feel free to take on the next chapter and dive in if you have the time. If another day's tasks end up taking you two hours, extend that chapter's work into the next day. It all evens out in the end. Flexibility is part of the agile approach.

When you're **sprint planning**, you're figuring out what input needs to go into a particular sprint. Don't worry—I've already made these decisions for you. Your sixty-minute task list every day for the next thirty days is done and waiting for you.

The **daily standup** is a self-assessment. Every day during a software project, team members keep each other accountable by asking each other short questions like, "What did you get done today? How are you doing since yesterday? What do you need to carry over to work on tomorrow?" At the end of each day (chapter) you'll see this simple multiple choice standup, and you'll check the appropriate box:

- ❏ I got today's tasks done early, so I'm going to get a head start on tomorrow.

- ❏ I got everything done today in about sixty minutes.

- ❏ I need to come back and do _____ tomorrow.

Will you need to check the first or third box often? It's doubtful. Your task list for each day includes only the essential activities that lead to your first, second, and third (and so on) paying customers.

A **sprint review** is exactly what it sounds like. At the end of each week, you review what you've accomplished. As I did with *The 60-Minute Startup*, this book will give you the opportunity to review what you got done during each sprint and to prepare for what's next.

In a **sprint retro**, you ask yourself how you can improve. What did you learn during the last sprint that you can apply in the future? I provide a space for you to answer this question after each week for your sprint retros.

I have a unique perspective on the agile approach because I used it both as a product manager, consultant, and entrepreneur. Before agile methodology, software development was tedious. We got detailed project requirements from the client in months one and two. We didn't start developing the product until the third month. Yet everybody knew the requirements would change. So when change occurred—priorities shifted, key personnel quit, the economy crashed, etc.—the first two months were wasted. When we took the agile approach, we went straight to prototyping in the first month. I'd show our client the first prototype of their user interface and logo, and they'd either say, "Yes, perfect," or "No, that's not what I meant." If they didn't like it, we went back to the drawing board. Either way, we lost no time.

This approach applies to all entrepreneurial journeys. When you draw up a business plan, it's never perfect, right? It's going to change because change happens. Not everything you write down on day one is going to stick. So you keep evolving that business plan. If you have a general idea about what you're going to do, that's good enough. Go do it, see what happens, and adjust your next steps.

This Book Will Work for You If . . .

. . . you pick an existing technology business. It's better to stick with a proven business than to create something from scratch that you have no idea will work out. In a coming chapter, we'll discover the type of business that's right for you. For now, I'd like to give you a preview of the many tech businesses that *could* work for you and your situation.

Graphic Design as a Service

While having a formal background in graphic design is absolutely going to be helpful, it's also relatively easy to learn the foundations of graphic design on your own. An increasingly easy-to-use Adobe Illustrator and websites like Canva and Visme are making it so that just about anyone with two opposable thumbs, a bit of creativity, and motivation can get paid to create or alter images.

Web Design

Web designers are incredibly valuable for technology companies. Web design is all about mastering the art of creating a beautiful, value-driven experience for the people using a website or app. There are always new websites popping up in need of professional web design, and companies like Skillcrush and General Assembly have proven online programs that'll get up-to-speed quickly with this career path.

Web Development

As a web developer, you'll build incredibly valuable skills that are in extremely high demand. You can get up to speed on building websites in as little as a few months with inexpensive or free online education programs like Treehouse and Codecademy. Once you command a knowledge of HTML, Ruby, Python, Javascript, or CSS, you can start taking on freelance projects to build your portfolio while you still hold your full-time job.

Online Courses

If you're an expert at something, there's likely an audience of people online who would be willing to pay to become an expert in your field—just like you.

Online Coaching

If you have something you're skilled at and very passionate about, you can turn that winning combination into offering your services with one-on-one online coaching.

Podcasting

If you can create a regular audience for your podcast on a specific topic, this is a great way to get sponsors. Other podcast related businesses include editing the audio of other podcasts and transcribing podcast episodes.

Virtual Assistant

Have a knack for staying organized? If you're a jack of all trades, you should consider working as a virtual assistant. You can find great gigs on Indeed or become a Zirtual Assistant. It can be an awesome way to rub shoulders with some very important people, build up your professional network, and you'll be able to work from anywhere.

Social Media Manager

We're all guilty of spending too much time on Facebook, Twitter, or Pinterest sometimes, so why not get paid to put your expertise to work? Lots of companies, especially startups or those in retail or travel, have a heavy social media presence, and are constantly in need of people to help build their brands online. You can find these types of opportunities on sites like Flexjobs and CareerBuilder.

Google Paid Ad Specialist

If you know a thing or two about paid internet marketing and are comfortable with Google, a great way to make some extra income on the side is to manage a company's Google Ad Campaigns, and gradually start bringing on more clients as your consulting business grows.

Blogging

Think blogging is no longer a viable source of income? Think again. Tens of thousands of bloggers, creating content on topics as diverse as scrapbooking, home cooking, travel, film, lifestyle, and more are launching self-employed careers thanks to a combination of blog subscribers, affiliate marketing, sponsorships, and other revenue streams.

Presentation Design Consultant

Yes, even the lowly PowerPoint requires outside consulting every now and then. I know I would happily outsource the visual layout of my presentation decks for work meetings, investor pitches, and lectures. Adam Noar from Presentation Panda is living proof that you can turn your passion for slideshow presentation design into a legit side business.

Landing Page Specialist

If you have a way with words and know how to make the keyword-friendly, beautifully designed, search engine optimized landing pages, why not charge other companies for your services? Even a short landing page is worth a couple hundred bucks in most cases. Take Freelance Copywriter Mike, who earns a great living writing landing page copy, autoresponder email sequences, and blog content.

Fiverr Gigs

Fiverr is a great place for first-time freelancers who might not have tons of experience and want to build up a portfolio of their work. You'll be able to complete simple tasks ranging from logo design, to creating animations, or even drawing a company logo on your forehead.

WordPress Website Consultant

Countless small businesses start out their web presence using a WordPress hosted website before needing to upgrade to other solutions for various reasons. Many of them will pay several hundreds of dollars for someone to get them set up online. If you have the patience to learn how to do it yourself, it's an extremely valuable skill.

Develop an App

Sometimes it seems like there's an app for everything. Yet somehow, new ones keep popping up and selling for lots of money, all the time. If you spot a niche that hasn't been filled to its potential just yet, and you can learn the coding skills (or know someone who already has them), you could be on to something. Just make sure you validate your app idea before jumping too far in.

Buy and Sell Domain Names

Domain name trading has been around for the last couple decades, and while the slam-dunk names have long been sold off (Insure.com went for $16 Million in 2009), there are still plenty of others that you can get your hands on for relatively cheap.[7] Imagine owning desirable domain names for the next decade's worth of innovative companies.

---

[7] Scott, Damien. "The 20 Most Expensive Domain Names Ever Sold." Complex, October 20, 2016. www.complex.com/pop-culture/2013/02/the-most-expensive-domain-names-ever-sold/insure.

Buy and Refurbish Used Electronics

Many people give up on their faulty laptops, mobile phones, or cameras without even looking into the cause of their malfunctions. If you have the skills to fix them, consider starting a side business refurbishing and reselling used electronics.

Data Analysis

Got a thing for numbers? Plenty of companies need to hire someone who's good at data analysis. Upwork and Digiserved are but two of many websites great for freelancers with an analytical prowess, looking for extra work.

Smartphone Repair

With so many consumers owning smartphones, there's a definite need for repair services. You can start a business where you have people bring in or send you their devices so you can fix various issues.

Refurbished Device Sales

Manufacturing actual computers and smartphones can be a larger scale operation. But you can still have a business where you sell those devices by refurbishing older models with new and improved parts.

Data Entry

Despite some "work from home" scams that involve tedious, low-paying data entry work, there are a lot of legitimate opportunities available for data entry businesses. If you are an exceptional typist with an eye for detail, explore what it takes to start a data entry business.

Internet Research

With the vast amount of information available online these days, being able to locate and validate information is a marketable skill. Do you have a knack for quickly finding relevant information on the Internet? If so, an Internet research business may be the perfect business idea for you.

IT Support

Many small businesses don't have the budget to afford a full-time, in-house technical support team, making outsourced information technology (IT) support a potentially lucrative business idea. You can provide network setup and maintenance services, data backup and recovery, and software installation and management. Depending on your clients' needs, your services can also include security measures and mobile device management.

Social Media Consulting

Social media is a powerful marketing tool, particularly for small businesses. If you are a social media expert and love helping others create their own social media presence, then you may be an excellent fit for a social media consulting business. Learn the pros and cons of starting a social media consulting business.

Video Production

Video marketing is growing in popularity for all types of businesses. Do you understand the technology behind video editing and enjoy the process of taking raw video footage and creating a video that flows? A video production business may be a great small business idea for you.

If your business idea relates in any way to the business types I've listed, *The 60-Minute Tech Startup* is for you. Because the agile approach is for you. To this day, I follow agile in every new entrepreneurial venture I start. I prototype, prove it works, tweak what doesn't, and evolve in the direction of getting customers. I'm going to teach you how to do the same, whether you're starting your first tech company or your next one. With this book, you will:

- Discover mistakes to avoid that doom most tech businesses

- Answer common questions such as "What is a non-compete?"

- Build a personal brand without getting in trouble at work

- Learn how to talk to your boss about your job on the side

- Perform a business model analysis

- Decide the type of tech business to start

- Borrow lessons of good side hustles that already exist

- Master customer outreach and get noticed

- Rock every sales discovery call

- Prepare for objection-handling

- Use the best ways to get referrals and grow organically

Really, Ramesh? All that in thirty days? you're probably thinking. Yes, every outcome I listed is not only possible, it's *likely* when you build your business the agile way. In the coming pages, I'm going to introduce you to fifteen agile tech entrepreneurs who started their businesses in record time. While aspiring entrepreneur friends sat around brainstorming business names, these entrepreneurs converted customers and earned referrals. Don't waste another minute on articles, emails, podcasts, tutorials, and lengthy business books that get you trapped in busy work. If you want to make money now, the agile way is the only way for you.

"But Ramesh, What If . . ."

With a promise like mine (paying customers in one hour a day in just sixty days) you might still have your doubts. I would if I were you! I bet you're thinking, *This agile business thing seems all well and good, but what if it doesn't work for me? I don't have any business experience.*

Fair question. The entire point of *The 60-Minute* approach is that you don't *need* any business experience. All the experience you need—knowing what to do and when to do it—comes with this book. You don't have to figure out anything by yourself. You're following in the footsteps of successful entrepreneurs who've built thriving tech companies the agile way. So the question isn't, "Will this work for me?" it's, "Can I follow instructions?" If you can, his book will work for you.

But Ramesh, what if I can't get everything done in just one hour a day? Won't it take like seven days to build my website, set up my social media, and write my marketing emails? Not with *The 60-Minute Tech Startup* it won't. Every day, I will give you what you need to get your tasks done. For example, when it's time to write your emails to reach out to prospects, I'll give you my templates. I've already tested these templates, so all you have to do is copy, paste, modify as necessary, and send. The same goes with every other complex task. Even drawing up contracts. The hard work is already done for you. If you can download a template, you can start your own business and get paying customers in thirty days or fewer.

Still worried you're the exception? That the agile approach I teach may not work in your corner of the high-tech industry, with your idea, or for you personally? *The 60-Minute Tech Startup* is for serious entrepreneurs who want a real tech company with real customers as soon as possible. Now, not every business model allows that. For example, if you're starting a data center, you need capital, server equipment, and employees. You can expect many months and lots of money to go into that business before you sign up your first account. I could not pass the red face test and promise you sales in thirty days. But if your dream is to open a data center, why not start a data management service where you work with a third-party provider? You won't need a lot of upfront commitment to start such an agile business. After you use this book to get your paying customers in thirty days, then you can start budgeting to open your own data center. First things first. It's the agile way.

No matter your background, idea, or existing skills, all you need to get started is this book, a computer, and an internet connection. A dedicated space such as a home office or a comfortable garage helps, but it's optional. If the laptop lifestyle appeals to you, you can build your own business from anywhere.

Did you know that 70 percent of people want to start a business but less than 15 percent of aspiring entrepreneurs go from idea to ownership?[8,9] My goal is to close this entrepreneurship desire gap and make you one of the few who starts, builds, and grows a viable, profitable tech company.

Let's begin.

---

[8] Reid, Susan. "The Numbers Are In: Most People Want To Be Their Own Boss." *Forbes Magazine*, January 8, 2016. www.forbes.com/sites/susanreid/2015/10/12/the-numbers-are-in-most-people-want-to-be-their-own-boss/#56f448171451.

[9] Buchanan, Leigh. "The U.S. Now Has 27 Million Entrepreneurs." Inc., September 2, 2015. www.inc.com/leigh-buchanan/us-entrepreneurship-reaches-record-highs.html.

60

Day 1 What Not to Do to Start a Tech Company

Is a Side Hustle the Best Thing for Your Day Job . . . or the Worst?

What is Big Data and what do people need to know about it? J. Cory Minton is on a mission to answer this question. Cory is the founder of The Big Data Beard, a creative media company authoring podcasts, blogs, and videos aimed at sharing knowledge about the innovations happening in Big Data, Machine Learning, Data Science, Artificial Intelligence, and the Internet of Things. Cory launched BigDataBeard.com on the side while employed full time at a large tech company as an engineer and tech evangelist. Even though he didn't think of it as a side hustle, BigDataBeard.com attracted a massive following over the years as well as paying sponsors. Currently, Big Data Beard is growing so fast that Cory is thinking of hiring employees to help manage the booming business.

It was Cory's day job that inspired him to start Big Data Beard. Six years ago, he wanted to share the creative work he was doing with other technologists, but his employer's corporate website wasn't the right place. So Cory decided to publish his own blog, talking about the interesting technical insights he'd gleaned from the industry and his personal views on where big data was headed next. As these opinion pieces got engagement, Cory found himself unable to keep up with readers' demand for more content. He reached out to several fellow engineers and offered his blog as an outlet for them to share their hands-on tech experiences. A handful accepted the invitation. Getting help with content opened up Cory's schedule to a new messaging medium—podcasting. But he soon realized Big Data Beard needed funding. He didn't want to cover podcast hosting services with his salary, and it just wasn't appropriate to ask his employer to pick up the tab.

Cory's first side hustle revenue came in the form of sponsorships. With over 160,000 podcast downloads, Cory had built up an audience that any tech company would want their message in front of. Since his decision to take Big Data Beard from fun side project to profitable side hustle, Cory has entered into sponsorships with Microsoft,

Disney, and book publisher O'Reilly Media. Now Cory is able to share meaningful interviews with tech influencers and what they have learned about technology with even more people.

Meanwhile, Cory continues to excel in his day job. Becoming a niche-famous podcaster and tech media influencer has polished Cory's communication skills and made him better at his job. Side hustle success like Big Data Beard will increase anyone's value to their employer! The business has meant the world to Cory's family as well. Literally, the world. He's been able to include his family on business trips to podcasting and tech conferences. His wife and their two young children have already visited nineteen countries and counting, including Australia and Japan. An unexpected benefit of these trips was another side hustle, this one started by Cory's wife. She now runs a travel blog where she shares travel tips with other families. When Cory and his family are back home in beautiful suburban Alabama, he drives race cars on weekends.

Cory's story seems like the side hustle gold standard. It didn't happen by accident. Cory had to be transparent about his side business, both with his employer and his family. Passion matters as does surrounding yourself with people who share it, as Cory did. But when it comes to starting a tech business on the side, knowing what *not* to do is essential to success.

Don't Start with Why, Start with Why Not

Why start an entrepreneurship guide telling you what not to do instead of focusing on what to do? Because of risk. I firmly believe that building a side business while you're still working full time is less risky, less stressful, and, later, less difficult should you decide to transition to your business full-time. To minimize risk and maximize income, you do not want to do *anything* that your employer might see as a conflict of interest or unethical. And that's why you should carve out clear boundaries around your side hustle and your day job. Otherwise, you may unknowingly blur the two, get in trouble, and lose your primary income before the side business is covering even your internet bill. So if you're not willing to keep your full-time job over the next thirty days (or longer), this book is not for you. Owning your own business probably isn't either.

That said, many successful entrepreneurs like Cory Minton have found a way to balance a job, a side business, and other responsibilities and obligations such as family. How did Cory do it, and how can you emulate what he did to make his business work from day one?

First, Cory told his employer he was thinking of starting a business. They loved the idea. Does that surprise you? It probably shouldn't. Employers love it when an employee wants to grow their skill set and not at the expense of the company. If you need to establish a personal brand to start your business and get your customers, you'll probably need to promote yourself online. And if so, your employer and colleagues will more than likely come to know about your side business. Like Cory, you are better off being proactive with your employer. If you decide to keep quiet about your business and take on clients by referral only, you may not need to tell your employer all about your business yet. Still, do the right thing. Cory did. He was open with his company about the fact that he would keep his side business activities completely separate from work even to the point of never, ever using company computers, hardware, or software for Big Data Beard.

Should you or should you not communicate with your employer about your side business? It's a tricky one. And completely your call. Cory believes in complete transparency. In his case, the employers also saw a benefit to his tech evangelism helping their business as well. In some cases, employers may not agree with it and may want you to choose. If you position it appropriately, there is a higher likelihood, especially nowadays when 39 percent of millennials have side jobs, your employer may not have any objections.[10]

Second, Cory did as he promised. He drew lines around his day job and his side gig and never blurred the two. He spoke with his employer's legal team about what he was up to so they could confirm there was no conflict of interest, accidental theft of intellectual property (IP), or information

---

[10] Morad, Renee. "Survey: More Than One-Third Of Working Millennials Have A Side Job." *Forbes Magazine*, October 1, 2016. www.forbes.com/sites/reneemorad/2016/09/29/survey-more-than-one-third-of-working-millennials-have-a-side-job/#fb6078f132f8.

privacy violations. This may seem like a big, hairy, scary task for you, but remember that companies want their employees to succeed. As you use your entrepreneurial adventure to test new ideas, you can bring your learnings to your day job to help your employer grow, too. Everyone wins, and you have the advantage of earning *two* paychecks.

Third, **Cory used common sense**. His full-time job has him working with secure data every day. So Cory wasn't about to start a side hustle as a hacker. Or even his own data security company. Check your employee agreement for the non-compete clause so you know you're in the clear. For Cory, clear boundaries means he will never sell Big Data Beard to somebody who would also buy his employer. But because Cory's side hustle benefits both parties, he's been open and honest, and he's kept the two businesses separate, he's down the right path.

In my experience launching, growing, and selling multiple companies while I worked full time, I've learned a few other not-to-do's in addition to Cory's. These must-avoid mistakes include:

Do not neglect your primary job. As a tech entrepreneur bootstrapping an agile business, your goal is to ease into a second income with minimal risk. That means keeping your main job for now. To do that, you must live up to your boss' expectations if not go above and beyond even when you're putting in ten to fifteen hours a week on the side working with clients. First things first!

Make sure your side hustle does not conflict with your main job. Cory's story covers this, but I've seen instances where entrepreneurs thought they were doing everything above board but were forced to choose between their job and their side business. Other conflict of interest examples include working with your employer's competitors, using their IP, competing with them, selling to their customers. If you're not sure what conflict looks like for your business, I advise you to seek legal counsel to answer your questions. You'll save yourself a lot of headaches and financial troubles later on.

Do not conduct your side business during work hours. This is fairly obvious but no less important. When you're at work, you're at work. No phone calls, emails, or direct messages to clients when you're in the office or clocked in. As the old boss' saying goes, "not on my time, not on my dime."

What Not to Do: Now It's Your Turn

In my first book in The 60-Minute Startup series, *The 60-Minute Startup*, I asked aspiring agile entrepreneurs to dedicate day one to documenting their knowledge, skills, assets, and attributes. We're taking a different approach with this book because chances are you know exactly what your knowledge, skills, assets, and attributes are. You're probably getting paid to use them in your day job in tech right now! Because tech skills transfer well from one business to another, it's more than likely that you'll be able to get paying customers on your own, but obviously not in competition with your employer. (If you want out of tech completely, it may be wiser to consult *The 60-Minute Startup* than this book, as daily activities are all about starting a profitable business when you have next to no clue what the business should do or sell.)

On your to-do list for day one is to complete the following checklist so you start right. Good luck!

❏ 1. Complete the What Not to Do Checklist: 60 minutes

Check off the box beside each task as you complete it.

- ❏ Did you talk to your employer or plan to talk to your employer about your side business?

- ❏ Did you make sure that you will not use your employer's assets (e.g., computer, phone, etc.) for your side business?

- ❏ Did you make sure that you will not use your employer's confidential information, even accidentally?

- ❏ Did you make sure there are no conflicts of interest such as working with your employer's competitor?

- ❏ Did you make sure that your side business hours do not overlap with your main job hours?

Daily Standup

Did you complete today's tasks?

- ❏ Yes

- ❏ No

If no, what do you need to carry over to work on tomorrow?

_____

What did you learn about your business (or yourself) today that will serve you in the future?

_____

Day 2 Choosing Your Tech Business

A Diamond in the Rough . . . Now on the Blockchain

You've probably know a thing or two about cryptocurrencies like Bitcoin. You may even have your own digital wallet where you store crypto you've bought. What gives most cryptocurrencies economic value is their real-world purpose. For example, Bitcoins are not really coins. They're "blocks" of ultra-secure data that allow anyone to see where that data is on the "chain." Bitcoin is the world's first decentralized peer-to-peer payment network and is powered by users with no central authority such as a government treasury or middleman such as a bank or credit card company. Here's a brief overview of how the technology works from Bitcoin.org.

> The block chain is a shared public ledger on which the entire Bitcoin network relies. All confirmed transactions are included in the block chain. It allows Bitcoin wallets to calculate their spendable balance so that new transactions can be verified thereby ensuring they're actually owned by the spender.[11]

Sukhi Jutla combined her blockchain developer skills, investment banking background, and jewelry industry knowledge to start a tech business unlike any the world has ever seen. Kind of like Bitcoin itself! Sukhi Jutla is a co-founder of MarketOrders, an independent retail marketplace helping jewellers get the products they need faster, cheaper, and directly from global suppliers. She is a sought-after international speaker and qualified IBM Blockchain Developer and has won numerous awards including Asian Women of Achievement, Female Entrepreneur of The Year, and Top 100 European Digital Pioneer by *The Financial Times* and Google. In 2018, Sukhi made global headlines when she became the world's first #1 bestselling Blockchain author.

Sukhi worked in investment banking for over a decade prior to starting up MarketOrders. In her role as a business analyst and project manager, she researched the jewelry industry

---

[11] "How Does Bitcoin Work?" Bitcoin. Accessed April 7, 2020. www.bitcoin.org/en/how-it-works.

and found that less than 4 percent of all transactions were done online simply because there was no secure way to process such transactions. That's what gave Sukhi the idea to start a tech company that seamlessly and securely digitized the jewelry supply chain so retailers and suppliers could access the gold and diamonds they needed without worrying about fraud. By using the blockchain to process transactions, Sukhi offers her clients both transparency and privacy, which no other technology besides crypto can. Previously, it typically took jewelers in the United Kingdom four months to receive orders from Singapore where most suppliers are. Now, MarketOrders is Uber for gold and diamonds, turning an old-fashioned way of doing business into one that takes a few minutes and a couple of clicks.

It all started small. Sukhi set up MarketOrders in less than ten minutes for £13, about $17. She then used her savings to outsource her website development to a company in India. Trialing her website for four months and generating over £60,000 ($78,000) in revenue validated Sukhi's business as viable. This confirmation gave her the courage to seek growth funding. Like most tech entrepreneurs, she sought out venture capitalists, and like most, her attempts failed. Even though she managed to secure one VC deal, Sukhi turned it down due to unfavorable terms. So Sukhi turned to the crowdfunding platform Crowdcube. After six months of crowdfunding, she soared past her £400,000 target goal, reaching £439,000 from 191 individual investors. That's well over half a million dollars!

At this writing, Sukhi's is scaling MarketOrders to serve even more clients and building a strong team to meet their needs. What amazes me about Sukhi's success is her seamless story. She didn't start a tech company based on a business she read about online or "what's hot right now." She simply repackaged her background and skill set in a new form—an online jewelry marketplace built on the blockchain. She didn't need to discover a great business idea, she *was* the business idea! So it can be with you and your tech company.

Which Business Is Best?

If you're thinking about starting a tech company, I assume you are somewhat familiar with technology. More than likely, you're adept at using technology. Perhaps you're a developer like Sukhi, a programmer, a graphic designer, or even a tech writer and transcriber. You have a technical skill set. You've found a home in one corner of the tech industry or another. Now you're ready to, like Sukhi, convert your knowledge and talents into a second income on the side. But which business is right for you?

Based on my experience as a serial tech entrepreneur and knowledge gleaned from interviewing hundreds of technology entrepreneurs for The Agile Entrepreneur, I've pinpointed ten tech side businesses with the highest chances of success, even if you've never started a company before. I've laid out these top ten tech side hustles alongside the knowledge and skills you need to make it. If your expertise matches one of these, that may be a business you should research further. If not, don't be discouraged. You can take tech skill-building courses, research other tech businesses beyond these top ten that correspond to your background, or even find a co-founder who can bring the necessary tech skills. For an expanded list of all side hustles, head on over to www.The60MinuteStartup.com.

Business	Required Knowledge and Skills
Web development and design	Familiarity with a content management system like Wordpress, Shopify, etc.
SEO consulting	Familiarity with search engine tools and knowledge of how search engines work
Tech blogging	Expertise in the domain (or access to experts) and knowledge about starting a blog
Tutoring and online courses	Expertise in at least one technology domain and ability to teach that expertise
Tech service freelancing	Expertise in the area of service (e.g.: coding, design, etc.)
Social media management	Expertise in driving engagement on popular social media platforms
Smartphone app development	Expertise in mobile application development for Android & iOS platforms
Graphic design	Expertise in applications such as Photoshop and knowledge in design concepts
Tech writing	Ability to understand technology and communicate via written media
Programming and quality assurance (QA) testing	Expertise in one or more programming languages such as Java, C, C+, and Python

What Business Should You Start?: Your Turn

Your goal for today is to match your knowledge, skills, and abilities to a viable business that you can call your own. To do that, pick a business from the list above, and list your areas of expertise that will make business happen. Here are your specific instructions.

❑ 1. Match your skills to your proposed tech business: 45 minutes

Download the **Required Business Skills** template for the above table at www.The60MinuteStartup.com. For this and all future templates, you can print it out and handwrite or just keep it saved on your computer and type into the fields. Whichever you choose, make sure you keep all your documents in one place so they're easy to find.

In the first column of the Required Business Skills template, list two to three potential tech businesses you could start based on your existing knowledge and skills. In the second column, write down your experience, knowledge, and skills that potentially match that business. As is true throughout this book, no answer is perfect. Nor does it need to be. You can always come back and tweak. The whole point of becoming an Agile Entrepreneur is to take action right away, learn, adjust, and move forward.

❏ 2. Identify the best-match business for you: 15 minutes

Now that you've looked at possible businesses to start, take a few minutes today to assess which business you might be most comfortable with, passionate about, and most likely to make your first sale within the next month. There is no science to this exercise. Remember, you can always come back to what you've written and change businesses if you later learn the first idea you ran with doesn't receive the market feedback (i.e., sales) you hoped. In fact, we're dedicating day five to getting that feedback before you invest too much time and effort. It's the agile way.

Daily Standup

Did you complete today's tasks?

❏ Yes

❏ No

If no, what do you need to carry over to work on tomorrow?

_____

What did you learn about your business (or yourself) today that will serve you in the future?

_____

60

Day 3 Other Profitable Skills Your Tech Startup Needs

Science on the Side

Can someone who works in a lab also work on the side? Kyle Isaacson is proving every day that the answer is yes. By day, Kyle is a nanotechnology, materials science, and chemical formulations researcher at the University of Utah. By night (and early mornings and weekends and holidays), Kyle runs Ike Scientific, a science consulting firm that draws on his expertise in the lab to help businesses develop new products, refine current product lines, and enter new technological spaces. Many small to midsize companies have a need for Kyle's skills, specifically those related to chemistry, but aren't willing to hire someone full-time. Unlike larger consulting firms that focus almost exclusively on high-level executive business consulting, Ike Scientific works directly with R&D and manufacturing teams. Ike Scientific does not require long-term contracts but works with clients on an as-needed basis. Kyle's services range from two-hour consulting sessions to year-long product development projects.

Ike Scientific is a young business, founded in April 2019. However, Kyle's side hustle turned a profit almost from day one because Ike Scientific offers low-overhead services rather than products that have to be developed, manufactured, or sourced. Kyle's success earned him the attention of *Utah Business Magazine*, where he was named to their "20 in their 20s" list.

Kyle's inspiration to start a tech company on the side came from a surprising course. At a financial self-reliance course offered at his church, Kyle learned about the importance of looking for other sources of income. Around that time, his academic research had begun to bore him. As he was in the last year of his PhD program, Kyle decided to see if he could leverage nine years of schooling into a lucrative gig. On a whim, he posted his services to a freelance website. Other scientists offered consulting services for as low as $20 an hour. Kyle wanted to stand out, so he set his price at $500 an hour! Remarkably, he received two consulting gig requests on the first day! These and other early gigs were one- or two-hour speed sessions, but several clients developed into long-term contracts. Still,

Kyle only works about ten billable hours per month. That equates to a $5,000 a month tech side hustle, and the demand for Kyle's services are only increasing.

To make Ike Scientific a viable business, he needed to develop complementary skills related to running a business such as accounting, legal, marketing, operations, invoicing, branding, design, and client relations. For example, after the first few consulting agreements, Kyle decided his appearance needed to match his high hourly rate. He quickly set up an LLC, purchased a domain name, and set up a simple one-page website. Recently, Kyle has been learning how to balance multiple projects at once. His clients are demanding. Persuading them to work around his schedule is difficult given that he still works full-time at the university in addition to his graduate studies. Often they expect Kyle to quickly solve problems they have been working on for months or even years. As he lives in a small apartment and cannot use his university lab for non-work-related activities, Kyle doesn't have the ability to perform chemical reactions and other scientific experiments. Therefore, he does all consultations via phone and video chat.

Despite the practical challenges and steep learning curve, Kyle is only getting started. He no longer needs freelance websites to sustain Ike Scientific because his website generates enough leads. Kyle only wishes he would have started sooner. If he had, Kyle feels that his company could now be valued around $10 million. If Kyle had started earlier, he would have easily had enough money to buy a house and rent a lab. Then again, Kyle is glad he didn't rush. He took his time researching the additional skills he needed to run a real business, and he learned them by doing them. For example, he watched website setup tutorials as he was building his site. This stretched Kyle's already limited time, but he had planned skills-building time into his schedule. The financial rewards made it all worth it.

The Most Important Skills You Need to Get Customers

More than likely, you're a solopreneur. You don't have co-founders, a core team, or investors backing you. Yesterday's exercise helped you identify the knowledge and skills you can bring right into your new business. But are they enough?

They weren't for Kyle, and yours probably aren't sufficient either. If you're going to get customers, you need to know sales and marketing. If you're going to be profitable, you need to know finance and accounting. Not to mention social media, project management, and website administration. I'm not saying you need to master these additional skills, right off the bat, but you should be familiar with the other skills you need and plan time into your schedule to learn them. That said, some areas might be better to outsource. Accounting is one example. You can delegate your bookkeeping somewhat easily, although I advise you to be familiar with accounting.

So which skills should you learn, and which should you outsource? The question is, how many hats can you wear? At a minimum, you should be able to pitch your business to potential customers, offer customer support, invoice clients (and get paid), and manage company expenses, especially during the first thirty days. Social media marketing is also a highly desirable skill to learn when you're looking for customers. I recommend you also consider soft skills for long-term success, such as professional communication, resourcefulness, patience, resilience, and management.

To plan for additional high-value complementary skills, use this matrix. You can find a downloadable and printable version at www.The60MinuteStartup.com. Here is an example for a logo design business.

	Yes, I already know	No, I need to learn or outsource
Specific domain knowledge and experience (e.g., programming, graphic design etc.)	I know Adobe Illustrator and Photoshop very well.	
Customer outreach, engagement, and support	I have engaged with customers in my job and am confident I can do this myself.	
Financial (business model, revenues, costs, etc.)		I'll learn this so it can help me in the long run.
Accounting		I'd like to outsource to an accountant I know.
Website		I don't need it initially, as I will use freelancing sites and Facebook groups to get initial customers. Will outsource later on.
Social media marketing	I will use Facebook groups and Instagram initially.	

If you decide to grow your side hustle into a full-fledged company with employees or contractors, you must upgrade your skill set to meet the demands of a growing business. You probably don't need to know how to vet, hire, and train a team right now, during the first thirty days, or even as you work with your first several clients. But I do want you equipped with the knowledge of the founder you'll need to be should you desire to scale this thing.

Harvard Business Review recently asked 141 Harvard Business School alumni founders, most of whom lead venture capital–backed technology startups, what knowledge and skills someone who aspires to their role needs. Here are their answers.[12] Consider these your long-term "to-learn" list!

What Skills Should an Aspiring Founder Prioritize?

% FOUNDERS WHO SAY THAT "HIGH" OR "VERY HIGH" PRIORITY SHOULD BE GIVEN

Skill	%
Founding team assembly e.g., choosing cofounders; splitting equity; recruiting advisors; managing a board	88%
Leadership e.g., setting and communicating vision; managing culture, ethical dilemmas, and cross-functional conflict	82
Product management e.g., prioritizing features; A/B testing; working with engineering and other functions	80
Team management e.g., hiring/firing; setting goals; giving feedback; designing organizational structure	74
Selling e.g., how to gauge interest and close; buyer's journey analysis; qualifying leads; sales compensation	72
Marketing e.g., sizing the market; product positioning; pricing; conversion rate optimization	71
Product design e.g., using personas and prototypes; UI design; usability testing; onboarding	69
Strategy formulation e.g., business model design; build/buy decisions; when to accelerate growth; when/how to diversify	65
Finance e.g., unit economics; cash flow forecasting; value added of investor types; VC deal terms; fundraising best practices	47
Engineering management e.g., managing technical debt; managing IP; ensuring scalability	44

SOURCE SURVEY OF 141 HARVARD BUSINESS SCHOOL ALUMNI WHO HAVE FOUNDED COMPANIES   © HBR.ORG

---

[12] Eisenmann, Thomas R., Rob Howe, and Beth Altringer. "What Does an Aspiring Founder Need to Know?" *Harvard Business Review*, August 1, 2018. www.hbr.org/2017/06/what-does-an-aspiring-founder-need-to-know.

Must-Have Skills: Now It's Your Turn

Now you know the most critical skills you need to launch your business and get paying customers in the first thirty days. Take some time to complete the skills matrix to determine if you have the skills or you will rely on external help.

❏ **1. Complete the skills possessed and needed matrix: 30 minutes**

Download the blank template or prepare a blank table on a piece of paper. Start with the minimum six rows shown in the example and add more as you see relevant for your business. Complete the table to the extent you can.

❏ **2. Plan to learn those skills or choose to outsource: 30 minutes**

For each of the areas you identified as must-learn or must-outsource, schedule into your calendar when you'll learn or hire out those skills. Where can you learn from a reputable instructor? YouTube? Udemy? A low-cost university alternative such as www.EdX.org? If you're going to delegate an area to a professional, who do you know who is an expert? If you can't think of anyone, google the skill you're looking for, add "near me" to your keyword search, and check out the highest rated and reviewed professionals who show up in search results. Reach out, let the business know your situation, and see how they can help!

Daily Standup

Did you complete today's tasks?

❏ Yes

❏ No

If no, what do you need to carry over to work on tomorrow?

_____

What did you learn about your business (or yourself) today that will serve you in the future?

_____

6

Day 4 Find an Unfulfilled Need, and You Found a Customer

The Surprising Place to Look for Your First Customer—You

Nathan Miller is President and CEO of Rentec Direct, a popular property management software company that helps property managers and landlords with day-to-day tasks. As a software, Rentec is like Quickbooks for landlords with features including an accounting system, online rent payment, and tenant screening. Because Nathan is also a real estate investor and developer himself, he had a need for streamlining project management long before starting Rentec. Nathan was a landlord managing five investment properties and wanted to own more, but working as a software developer full-time while property managing took up his free time. The only way he could see himself growing his real estate investment portfolio was if he could somehow automate the most tedious property management tasks.

So Nathan googled around for property management software to make landlording easier but found nothing suitable. That's when he decided to build his own software, Rentec Direct. Nathan considered it a hobby, not a product, so he shared the platform with other landlords for free. They loved it. Software took the frustration out of recurring tedious tasks like finding and vetting new tenants. No competing product was as easy to use as Rentec Direct.

But Nathan's first users wanted more features. And so did Nathan. These demands to enhance the software motivated him to turn Rentec Direct into a paid subscription model. Nathan hired his first employee and was officially in business. His first revenue came from the landlords Nathan had given free access to, and the organic growth continued. Through blog articles Nathan wrote, Rentec Direct attained the number-one spot on Google for landlord property management software. The business continues to grow to this day through customer referrals, and in large part because the software was built first and foremost to fulfill a need Nathan knew existed.

Your Customer: Who, Where, and How

Two of the most pressing questions for an aspiring entrepreneur are:

1. What do I sell?

2. Who do I sell to?

Days one through three answered the first question. Let's now tackle your customer base, the people you're going to sell your product or service to. Remember, our primary goal is to get one or more paying customers in the first thirty days of your entrepreneurial journey. Believe me, these initial customers will give you the confidence (and valuable lessons) you need to build the business long-term.

Today, our task is to identify your future customers by their most pressing pain points. As host of The Agile Entrepreneur podcast and speaker at many entrepreneurship events, I've had the good fortune of asking hundreds of successful entrepreneurs how they found their first customer. About eight of every ten people I ask say their first customer was either someone they knew or a person referred by someone they know. Hence, I strongly recommend you talk to your family, friends, colleagues, and neighbors about your business. Don't pitch them anything yet, though. I just want you to feel comfortable talking about your business idea and what you plan to offer. You may find people saying what Nathan's landlord friends told him when they learned about his property management software.

"Wow. That's exactly what I need. Can you let me use it, too?"

If you can't see yourself selling your product or service to anyone in your immediate network (or anyone in their networks), there are other ways to research potential customers and their pain points. Freelance platforms such as Upwork, Fiverr, Toptal, Simplyhired, PeoplePerHour, and Guru are extremely useful for customer research. If I were planning on a logo design business, I'd go on these websites, search for logo design jobs, see what clients are asking for, and note what top-ranked professionals are offering.

Another customer research method is keyword research. Every day, your future customers are googling potential solutions for their pain points. You can tap into the power of search to find out the exact words customers use to describe their pain points. My top free keyword research services are Google Adwords Keyword planner, Wordtracker, SEMRush, and Answer the Public. Because each platform has its own tools that take time to learn, I've created tutorials for each keyword research service so you can uncover your customers' desires faster than ever. Access these free tutorials at www.The60MinuteStartup.com.

Whichever way you use to research your customers—talking to potential buyers, checking out freelance platforms, using keyword research, or all three—complete the following table. I use the logo design business as an example.

Potential customer	Pain points	Where can I find them?
Friends	Can't find a logo designer they can interact with on a personal basis	In your friends' circle
Small business users using Upwork.com	Looking for inexpensive but fast logo design	Post a job on Upwork.com

As an aside, pain points typically fall into four categories: financial, productivity, process, and support. Customers with a financial pain point want to make more or save more money. A productivity pain point drives people to find ways to get more things done in less time. People with a process pain point need help going from point A to point Z without skipping critical steps. And customers with support pain points simply need a third party's help planning and executing. Either they don't have the knowledge, abilities, or bandwidth to DIY a solution, or they need an expert guide to tell them what to do to achieve the intended result.

If you'd like a deeper exploration of the four pain point categories and which is more profitable to address, refer back to book one in this series, *The 60-Minute Startup*.

Who Is Your Customer?: Now It's Your Turn

Download the customer pain point template from www.The60MinuteStartup.com or create your own along the lines I have shown you.

❑ **1. List potential family, friends, and colleagues you can talk to: 15 minutes**

Today's exercise is not about actually talking to your network but only listing those you could potentially talk to about your business. As a reminder, you're not going to ask them to buy. You will simply let them know about your business plans.

❑ **2. Do online research using freelancing services platforms: 30 minutes**

Visits websites like Upwork.com, Fiverr.com, Toptal.com and identify potential customers and their pain points based on real clients' job postings and descriptions.

❑ **3. Do additional research using at least one keyword research tool: 15 minutes**

In the remaining time today, complete additional research using one of the keyword research tools I recommend, such as SEMRush.

Daily Standup

Did you complete today's tasks?

- ❏ Yes

- ❏ No

If no, what do you need to carry over to work on tomorrow?

_____

What did you learn about your business (or yourself) today that will serve you in the future?

_____

Day 5 Show 'n Tell to Impress Prospects

O Canada . . . Here I Come

If you haven't heard of Vartika Manasvi before reading this book, chances are you'll be hearing a lot about her soon. Vartika's claim to fame is her story of risk. She bought a one-way airplane ticket to Canada with a dream—building a tech company. Her dream came true. Vartika is the first South Asian female entrepreneur to receive a Canadian startup visa. Her business StackRaft is like LinkedIn for software engineers. Professionals can network, meet mentors, solve common career challenges, upgrade their industry knowledge, get skills training, and apply for jobs . . . all without that "sponsored message" spam.

Like every social network founder before her, Vartika had to find a way to make money off her platform. For StackRaft, that means charging companies seeking software engineers for a curated list of top candidates based on their skills as verified by StackRaft. But unlike other social media websites, StackRaft didn't focus on acquiring as many users as possible, at least not at first. To lower the risk of tech startup failure, Vartika decided to build a basic proof of concept she could pitch to tech companies where she lived in Toronto. Vartika was new to the city, having no friends or network. But she understood those companies' pain points, which are recruiting, vetting, interviewing, and hiring software engineers. Especially difficult are validating a candidate's skills, personality, and career intent. StackRaft would offer data insights from digital footprints combined with unique skill-based challenges to help companies find the perfect fit.

Vartika also knew there was demand from software engineers for an idea like hers. Being a first generation entrepreneur, she had over one hundred conversations with fellow immigrants to Canada who needed help finding jobs in the tech industry.

Vartika didn't need to build the StackRaft platform or even finish the website to show off its value. That proof of concept brought Vartika her first paying customer. She incorporated the company the next day and delivered for that first customer. She then teamed up with a cofounder to invest a year in building the full-fledged StackRaft platform while also acquiring customers and users. Today, StackRaft is the world's best place for small to medium-sized companies with remote and distributed teams to hire global engineering talent.

What Do You Need For a Show 'N Tell?

Like Vartika, all you need to get a green light for your business idea is a proof of concept. Some call this a minimum viable product, a prototype, an idea validation vehicle, a concept vehicle, or wireframe. Whatever term you prefer, the point is, you're investing bare-minimum time, effort, and money into your business idea so you have *something* to validate your idea with potential customers, get their feedback, generate excitement, and even earn a commitment to buy your product or service when it's ready. This will give you the confidence (possibly even the cash!) to work on and finish your offer.

So, what does a show 'n tell vehicle look like? What are you going to show these prospective customers? Well, that depends on your particular type of tech business. If you're developing a software application, your show 'n tell could be a simple wireframe of the final product. If you're working on an online e-commerce business, you can start with a rough outline or sketch of your website and highlight how your value proposition, online shopping experience, product selection, or price are superior to existing companies. If your side hustle is graphic design, assemble a portfolio of previous designs to show companies or individuals you'd like to work with in the future. If you're a technical writer, start with content samples. You get the idea.

The simplest way to package your show 'n tell is, believe it or not, a slideshow. That's exactly what I did to create a proof of concept for my data science and management consulting businesses. I designed a PowerPoint slideshow, recorded a webinar, and sent it to several prospects to generate interest. It worked, and both of those businesses are thriving to this day. Of course, a slideshow is not the only way. It may not be ideal for content like writing samples, which are better sent in word processing documents or PDF files. Here are a few more ideas and tools to make your business idea come to life before you actually create it:

- A video demonstration
  (Check out www.Camtasia.com, a paid software tool that allows you to produce professional audiovisual content.)

- Infographics
  (Check out www.Canva.com, an easy image design platform especially helpful if you've never done graphic design.)

- Wireframes for apps, websites, or landing pages (Check out online wireframe design platforms such as Sketch.com, Adobe XD, Figma, Invision Studio, Gliffy.com, LovelyCharts.com, NinjaMock.com, and Mockflow.com. For a complete list, head over to www.The60MinuteStartup.com.)

- An online shareable slideshow
  (Check out www.Slideshare.com, a LinkedIn-owned
  platform for designing engaging presentations.)

The possibilities are endless!

Your Show 'N Tell: Now It's Your Turn

Yesterday, you identified your customers' pain points. Now let's think through how you can demonstrate the fact that your product or service can resolve those problems. Also, how will you get your proof of concept in front of these prospects for your show 'n tell? Let's get to work.

❏ **1. Figure out how you'll share your show 'n tell: 15 minutes**

Will you meet with prospects face-to-face? Or are you going to show them remotely? Email your portfolio, slideshow, or video? Or will you mail physical objects such as a folder of previous designs?

❏ **2. Create your show 'n tell: 45 minutes**

Whether you're recording a video, designing an infographic, or building a wireframe, it's perfectly acceptable if it takes longer than forty-five minutes. In fact, that's probably the minimum you should spend on your show 'n tell vehicle. As I've said before, the sixty-minute concept is a loose framework. Some tasks take longer, and that's OK. What matters most is that you're taking *only* the necessary steps that lead you towards paying customers. That's what today is all about. Rubber, meet road!

Daily Standup

Did you complete today's tasks?

❏ Yes

❏ No

If no, what do you need to carry over to work on tomorrow?

_____

What did you learn about your business (or yourself) today that will serve you in the future?

_____

6

Day 6 Intellectual Property: Knowledge Is Power

A Smart Way to Earn Passive Income

If you're curious about passive income, you have probably heard of Pat Flynn, creator of www.SmartPassiveIncome.com. After Pat got laid off from his job in May 2008, he started a blog to earn passive income online and pay bills. Google AdSense affiliating was his first revenue stream followed by an ebook on how to ace the Leadership in Energy and Environmental Design (LEED) exam. Pat's background is as an architect, so he was writing about what he knew. He himself had to study for the exam, which was not an easy task. Most guides Pat found at the time were disorganized and difficult to memorize. That's what led him to write the guide he wished he would have had. He was on to something! In a few short months, Pat had amassed over three million page views![13]

Pat's success with that ebook— www.GreenExamAcademy.com—appeared short-lived when he received a cease-and-desist letter. The original domain name of Pat's test prep website was www.InTheLEED.com, but "LEED" was trademarked. After weighing pros (very few) and cons (very many), Pat and his lawyer decided to pivot, and Green Exam Academy was (re)born.

Today, Pat authors bestselling books on how to launch products that generate passive income, sells courses on how to build an online business, and releases free tools for entrepreneurs that smooth out daily business operations. All this started with those simple affiliate links and an ebook. With a wife and two children he can spend most of the day with, Pat is truly living the entrepreneurial dream. I think even Pat would admit that the nightmare of a cease-and-desist letter helped him get where he is today. Fortunately, Pat's struggle means that you won't have to

---

[13] Flynn, Patt. "The History of My First Online Business." Smart Passive Income, April 25, 2012. www.smartpassiveincome.com/my-first-online-business/.

endure the same. That's why today is all about the law—and what you need to know to stay within it.

Everything Side Hustlers Need to Know about Intellectual Property (IP)

I am not a lawyer, so I am not pretending to be one. Please consult a lawyer for all things legal, IP issues included. I'll share what I've learned and experienced in my own journey. The US Patent and Trademark Office (USPTO) offers a useful overview of the four types of IP:

1. Copyright: A form of protection offered to the authors of original works of authorship which protects literary, dramatic, musical, artistic, and certain other intellectual works.

2. Trademark: Any word, name, symbol, or device or any combination, used, or intended to be used, in commerce to identify and distinguish the goods or services. Also includes logos, banners, sound, smell etc.

3. Trade secrets: Any information that provides economic value that is not in the public domain and that has been reasonably kept secret. Formulas, software programs, techniques etc. fall into this category.

4. Patent: A grant of property rights by the US Government.[14]

---

[14] Purvis, Sue A. "The Fundamentals of Intellectual Property for the Entrepreneur." United States Patent and Trademark Office. Accessed April 7, 2020. www.uspto.gov/sites/default/files/about/offices/ous/121115.pdf.

To make sure your IP is protected, research your chosen business and domain name(s) online, which Pat Flynn admits he should have done. Are you planning on using names, terms, or words that are trademarked or copyrighted by another business? If so, do what Pat did— change them. Better safe than sorry.

After doing your research, file for IP protection for one of three things, whichever you want to protect:

1. Copyright

2. Trademarks

3. Patent

If you believe you have trade secrets, draft a Non-Disclosure Agreement (NDA) and ask anyone you talk to about your business idea, methods, or technology to sign it first. To further protect your IP, you might look into IP insurance. The USPTO has excellent resources on lawyers who provide this service, the cost of filing fees, and more.[15]

Protect Your IP: Now It's Your Turn

Is IP worth looking into? You may be wondering if your idea is worth protecting, just as I did when I authored *The 60-Minute Startup*. Believe me, this will be the best sixty minutes you'll spend today. Getting to know about IP and

---

[15] "Attorneys, Agents and Paralegals." United States Patent and Trademark Office, August 13, 2019. www.uspto.gov/learning-and-resources/attorneys-agents-and-paralegals.

any potential advantages or threats you might have in this space will save you from many headaches.

❏ 1. Getting to Know about IP: 20 minutes

Surf through the resources I've mentioned so you are fully aware of what IP is, what you need to protect, or where you may be potentially infringing. Be proactive about IP, not reactive.

❏ 2. List any Potential IP You May Have to Protect: 20 minutes

It may be a little early, but if you can, list your potential domain name, logo text, and graphics you are thinking of using for your business. Also list out any trade secrets like your programming algorithms or a custom procedure you've developed.

❏ 3. Assess Opportunities and Threats: 20 minutes

Now evaluate if you want to protect any of your creations with a copyright, trademark, or patent. Also evaluate if any of your assets might potentially infringe on others' IP.

Daily Standup

Did you complete today's tasks?

❏ Yes

❏ No

If no, what do you need to carry over to work on tomorrow?

_____

What did you learn about your business (or yourself) today that will serve you in the future?

_____

Day 7 Make It Official

A Cheeky Way to Find Your Dream Date

Lori Cheek is a NYC-based architect turned entrepreneur, and the Founder and CEO of www.Cheekd.com, a hyper-speed Bluetooth mobile dating app that removes the "missed" from "missed connections." She was listed as one of AlleyWatch's "20 most awesome people to know in the NYC tech scene" and has been called "The Digital Dating Disruptor."[16] After

---

[16] AlleyWatch. "20 Awesome People in the New York Tech Scene You Need to Know About." AlleyWatch, April 23, 2017. www.alleywatch.com/2014/08/20-awesome-people-in-the-new-york-tech-scene-you-need-to-know-about/21/.

working in architecture, furniture, and design for fifteen years, Lori abandoned her career for an idea that led her into the NYC world of technology and dating. Instead of building structures, she's now building relationships. So where did her inspiration come from?

One night, Lori was out to dinner with a friend and colleague. As they were leaving the restaurant, she saw her friend scribble something on the back of one of his business cards. Then he slid the card to an attractive woman at a nearby table as they walked past. Lori caught a glimpse of the scribbled message.

Want to have dinner?

Lori went home that night with an idea. A thousand times during her NYC commute, she'd spotted an intriguing stranger on a train, in a café, crossing the street, at baggage claim. You name it. All but a few got away. Handing out a business card could have been the answer, but the personal details included on a typical business card was simply too much information to hand to a total stranger. Online dating was intriguing, but what if that initial encounter could be in person?

That Saturday, Lori gathered a group of friends and many bottles of wine, and they spent the day brainstorming ideas. At the end of the day, Lori had a plan. People would express their interest in a romantic prospect by coyly handing them a small black card with a cheeky phrase, such as "Act natural. We can get awkward later," or "I just put all my drinks on your tab." This would work whether they were feeling awkward, shy, or simply wanted a new approach. The recipient of the card would then be invited to go online and check out the user's profile. Then they'd decide whether they wanted to make contact through a private messaging service, where personal information was protected.

But what would she call her company? Within five minutes of her girlfriend asking the question, Lori had the answer. She took her last name and added a "d." Her future customers were about to get "Cheekd."

Lori launched Cheekd in May 2010. A few months later, Cheekd popped up on the cover of the Styles Section of *The New York Times . . .* "Move over, Match.com, this is the next generation of online dating."[17] A couple of days later, Lori got a call from Oprah Winfrey's Studio asking for an interview. That's when Lori left her job and started working full time on Cheekd. Soon after, Cheekd went global with customers in forty-seven states in America and twenty-eight countries internationally.

But financial success was slow going. Finally, after four tumultuous years of building her startup with bad decisions and rookie mistakes, she applied to ABC's *Shark Tank* to take Cheekd to the next level.

When Lori proclaimed she was going to change the population with my reverse engineered online dating

---

[17] Rosenbloom, Stephanie. "The New Dating Tools: A Card and a Wink." *The New York Times,* July 21, 2010.
www.nytimes.com/2010/07/22/fashion/22date.html.

business, serial entrepreneur and Dallas Mavericks owner Mark Cuban rolled his eyes, called her delusional, and immediately snapped, "I'm out." Billionaire investor, Kevin O'Leary, demanded that she quit her "hobby" and shoot Cheekd like a rabid dog. After getting shot down by all five Sharks, Lori looked them in the eye and said, "Trust that you'll all see me again." Although those final bold words ended up on the cutting room floor, in the forty-eight hours after the broadcast, Cheekd.com received a record-breaking 100,000 unique visitors. Lori's inbox filled up with thousands of emails insisting that the sharks were "out of their minds" for not investing.

Nearly fifty of those emails were from other interested investors. Lori ended up raising five times the amount of money she'd sought and got a CTO on board who helped facilitate and finance the new technology behind Cheekd—one that wasn't available when the patented Cheekd idea was launched in 2010. While handing out the physical cards worked anywhere in the world and were a perfect way to break the ice, there were a few barriers; the main one being that many users were still quite intimidated to walk up to a total stranger. So Cheekd now uses a cross-platform, low-energy Bluetooth technology. You now get a notification if someone who meets your criteria is within thirty feet of you. If you're near a potential spark, Cheekd makes sure you know about it. The app connects people in real time. Unlike old online dating, connections begin in person—then Cheekd helps you take the next step and continue the conversation online.

Investing in a Good Name (And Making it Official)

So, how did Lori come up with the name Cheeked? She didn't spend time stressing out over what she'd call her business. And she didn't have to go too far in coming up with one. She simply started with her last name—and voilà! Business name, check. There is no "right" way to come up with a business name.

Next, Lori had to decide on a legal structure. Unfortunately, entrepreneurship in the digital age isn't like selling lemonade in your front yard as a kid. You have to complete and submit paperwork to the government so the authorities know your business exists. (Don't worry, we'll cover that—and it won't take hours and hours!)

Like Lori, you get to decide between a sole proprietorship, partnership, corporation, and limited liability company (LLC). Given that Cheekd involved thousands of peoples' personal information, Lori needed a lawyer who specialized in data privacy and cyber security. She researched attorneys and lawyers online, asked colleagues for referrals, and picked a lawyer to help protect her business as she grew. Lori then went through the pros and cons of the possible legal structures for her business. With her lawyer's guidance, she decided on an S-Corporation. For entrepreneurs who care about getting revenue as fast as possible (and keeping as much as possible), LLC makes the most sense. You'll soon learn why.

Business Name and Legal Structure: Now It's Your Turn

Check off the box beside each task as you complete it.

❏ 1. Select a business name: 10 minutes

Naming your business can feel like naming a child. It's not that important, but it sure feels that way. Over the years, I've met entrepreneurs who spent tens of thousands of dollars with branding consultants to come up with the perfect business name. Don't waste your money or your time! Why are you starting this business? To get customers. That's our single objective. So your goal for your business name should be selecting one that doesn't *repel* potential customers. That's it. My best advice is to pick whatever comes to mind when you think about the people you want to help and what you want to offer. It can be anything. This isn't a science.

Got a business name in your head? Great. Now, to make sure your name won't push customers away, follow these simple rules:

- Search your business name online to confirm it's not already taken

- Avoid hard-to-spell names (e.g., Acquaintance Inc.)

- Don't pick a name that limits your business as you grow (e.g., Only Pine Candles LLC)

- Use a name that gives people a clue what you do (e.g., Internet Marketers of Indiana)

- Keep it simple and short (e.g., Purely Sweet Baked Goods)
- Don't copycat other businesses (e.g., The North Face Book Printing Company)
- Stay away from random or annoying words (e.g., Farting Hounds Popcorn)
- Don't use industry jargon people don't know (e.g., Onomatopoeia Freelance Editing)

Does your business name idea pass the test? If not, run your next idea through the rules until it passes. If so, congratulations! You're ready to move on to step two.

❏ 2. Secure your domain: 10 minutes

You have a business name, and you know the matching domain name isn't taken (e.g., www.YourBusinessName.com). Now it's time to go buy it! Take ten minutes and go to www.GoDaddy.com right now to claim your name online. It should cost you no more than $15. That's it. That's all you have to do.

❏ 3. Register your business with the government: 30 minutes

If your business isn't registered with the government, it's not a real business. If an angry customer takes legal action against your business for any reason, you are held personally liable. They could sue you, seize your bank accounts, and take your house. Unlikely, but possible. It's hard to serve your paying customers if one of them gets angry and has you thrown out onto the street. Better safe than sorry—make your business a legal entity. What do I mean by legal entity? Well, you've seen the types of business structures already:

- Sole proprietorship
- Partnership
- Corporation
- Limited Liability Company

If you're not pursuing outside investors, teaming up with other people, or planning to sell shares of ownership in your company, go with an LLC. When your business becomes a limited liability company, you and your business are now separate entities. Disgruntled customers can send their lawyer sharks after your business, but not after you and your family. Thank goodness!

What about taxes? Businesses pay taxes just like employees do. In the United States, the Internal Revenue Service (IRS) automatically classifies an LLC as either a partnership or a sole proprietorship. If you're the sole owner of your business, your LLC is a sole proprietorship. That's the case for 99 percent of people reading this book.

All you need to file for an LLC is your business name and some cash. LLC state filing fees range between $50 and $500. The average filing fee for an LLC in the United States is $127. To download your LLC paperwork, file for an LLC, and pay the fee, go to www.The60MinuteStartup.com for the most up-to-date website links.

❏ 4. Find legal representation (in case you ever need it): 10 minutes

Filing the paperwork for your LLC is quick and easy, so you shouldn't need an attorney. However, smart entrepreneurs always have a general business attorney and an intellectual property (IP) attorney in their contacts, should they need them. A general business attorney will be able to answer any questions that come up around customer contracts, agreements, or terms and conditions. An IP attorney will help you protect anything you create for your business from theft by competition. If you don't know any attorneys already, the fastest way to find trustworthy attorneys is to Google "business attorney" and "IP attorney" followed by your country or state. Bookmark the business attorney and the IP attorney with the best reviews or add them to your contacts. You're all set for today.

Daily Standup

Did you complete today's tasks?

❏ Yes

❏ No

If no, what do you need to carry over to work on tomorrow?

_____

What did you learn about your business (or yourself) today that will serve you in the future?

_____

Day 8 Priced to Sell

To Sell Fish or Teach People to Catch One?

Will Roberts III and John Holmes II are the Executive Partners and founders of WeWorked.com, a bootstrapped and profitable provider of online timesheet software to businesses in over 120 countries worldwide.

Before founding WeWorked, John led large-scale IT acquisitions for the District of Columbia and Federal Government and successfully managed domestic and international multimillion-dollar software projects, spanning from the U.S. to Egypt. Prior to co-founding WeWorked, Will worked for and managed software development and technical support staff for federal contractors. He was also Bank of America's Chief Data Architect at their Business Loan Decisioning Center. Both John and Will's prior employment history served as major inspirations for what was to come next.

In 2009, John and Will noticed that small businesses were struggling with tracking their staff's hours. What if they could create a web-based timesheet software with features that focused on the needs of the small business? But building a web application outside of Silicon Valley is difficult today and was unheard of in 2009. In the Washington DC area, most software companies focus on government business by building expensive enterprise applications. Good, affordable small business software is rare.

They didn't get much support. People either thought they were crazy or ignored them all together. "Someone already does that," "It'll never work," and "You think you know how to build something like that?" were among the common responses. They had no blueprint, just a dream and persistence.

John and Will did have competitors, but they were large companies that couldn't survive thirty days without VC money. Still, it was hard to get noticed. In 2009, they launched WeWorked.com as a "freemium" product, meaning subscribers were allowed to use a full-featured version of the software at no charge. They hoped to gain feedback and make improvements. But it was apparent that small businesses were reluctant to trust their time data with any product that was being given away for free. So John and Will went back to the drawing board for a new marketing strategy. They gained market disruption through below-market pricing and aggressive advertising.

How could they price their software so low and still make a profit? WeWorked.com used some unique concepts. They offered email-only support with no contact number or telephone support. Unlike their competitors, they bundled their user licenses and didn't offer the traditional per user pricing model. Both of these concepts allowed them to offer lower pricing to small businesses. And ten years later, with more than 50,000 users in 120-plus countries, they have done the impossible. They're profitable and competing with big name timesheet companies with annual revenues of $20-$40M.

John and Will credit two highly influential mentors that guided them throughout the process. Their names are "Trial" and "Error"! Given that many of their concepts were groundbreaking, there was no frame of reference or advisers to seek guidance from. They literally tried different stuff and made adjustments based on the results. Every setback they encountered and overcame led them to where they are today.

The best business ideas don't require huge amounts of money to get started. Your time is your most valuable asset. Don't be afraid to launch a product that already exists. Sometimes there's an existing void within a market that can be tapped into very lucratively. Technology can be used to reach customers all over the world. The right product, pricing, and marketing can allow you to compete with the biggest and best.

How to Price Your Product or Service

Pricing is an important piece of the puzzle, and you probably won't get it right on your first try. I don't mean to sound like a downer, but any successful entrepreneur will tell you that correct pricing takes experimentation. The important thing is to use guidelines to get you as close to the mark as possible.

John and Will didn't get it right the first time. They iterated and eventually settled on something that felt right for them and their customers. To find your own sweet spot, you first need to decide if you'll charge hourly prices or flat rates (or some combination of the two). Here's a chart to help you:

	Pros	Cons	When to Use
Hourly Price	Popular, easily understood, earn for what you work	Caps earnings, punishes efficiency	New to freelancing, Ad-hoc work, Unfamiliar / unstructured jobs
Fixed Price	Simple, budget-friendly, maximize earnings	Risk underselling, inflexible, hard sell	Well-scoped work, short projects, budget-conscious clients, trusted relationship

Beyond that, there are five basic pricing strategies you can use. Here are their basics:

	What it is	When to use
Cost Plus pricing	Calculate all your costs and add your profit margin	Typically used for products when you can identify all costs
Competitive pricing	Set a price based on what your competition is charging	Use it if you know your competition and have a general idea of their prices
Value based pricing	Price based on what the customer perceives is the value of your offering	Somewhat difficult to figure out; use it if you have a unique offering
Price skimming	Start with premium price and reduce over time	Use it if you are the first in your category
Penetration pricing	Set a lower price to get a foothold and increase over time	Use it when you are new and don't have many referrals

Today's task is figuring out which pricing strategy you'll use and the number amount you'll start with. Yes, you can still get clients with imperfect pricing. Don't worry. You're still on track to have a paying client by day thirty.

Decide on Your First Pricing Strategy and Price Point

Check off the box beside each task as you complete it.

❏ **1. Spy on your competition: 20 minutes**

To get an idea of what your target market is used to, look up your competition's pricing. Do they charge hourly? Sell packaged deals? Do they price competitively? Or do their prices vary greatly from each other for the same services?

❏ **2. Choose a sensible pricing strategy: 30 minutes**

Are you offering a service or a product? What are your upfront costs to deliver your product or service? How much of a profit margin do you need to make? Do you want to get as many customers as possible, or do you want to be a premium leader in the market? Take some time to write down your answers and compare them with the pricing strategies in this chapter. Which one makes the most sense to start with?

❏ **3. Decide your price and justify it: 10 minutes**

Now that you know your pricing strategy, choose a price to start with. Remember, don't expect to stick to this price forever. It's a starting point. For today, write down the price point you'll start at and your reasons behind it. This way, when you do adjust your pricing, you can do it systematically and based on logic.

Daily Standup

Did you complete today's tasks?

❏ Yes

❏ No

If no, what do you need to carry over to work on tomorrow?

_____

What did you learn about your business (or yourself) today that will serve you in the future?

_____

6

Day 9: To Plan or Not to Plan?

Mint Your Way to Digital Success

Brian Meert knows a thing or two about being creative. He is founder and CEO of www.AdvertiseMint.com, an agency that helps businesses grow using Facebook advertisements. Like every entrepreneur profiled in this book, Brian started his company on the side. Back in 2014 when Facebook launched its own advertising management platform, Brian, then the Vice President of a financial services company, ran a few Facebook ads on behalf of his employer. This wasn't Brian's first brush with paid advertising. Brian started his first venture while earning his MBA. He accepted a challenge from the dean of the business school, using his campus cafeteria money to test Google AdWords for keywords like "goal setting" to drive web traffic. Even though the part-time company was not wildly successful, Brian learned how to start a business with very little money.

So when Brian saw immediate results with Facebook ads, he knew he had a real opportunity. He told his boss that he wanted to start his own advertising business. Would the financial services company be his first client? A bold move. And it worked. Brian's employer agreed, Brian launched www.AdvertiseMint.com, authored *The Complete Guide to Facebook Advertising*, started The Duke of Digital Podcast, and the rest is history.

Of course, it wasn't all hunky dory for Brian on his entrepreneurial journey though. Prior to Advertiisemint.com, Brian started another company, which taught him the importance of cash flow management and more importantly about planning.

Business Planning: Prepare One But Don't Dwell on It

I'm not a big fan of elaborate business plans. Many successful entrepreneurs I've interviewed never had a formal, complete business plan. Still, I believe everyone should plan how they're going to launch their business. Surveys show that businesses with a plan have a higher success rate than those without one.[18] Unless you're going for funding from the Small

---

[18] Lesonsky, Rieva. "A Business Plan Doubles Your Chances for Success, Says a New Survey." Small Business Trends, January 20, 2016.

Business Association or some other organization, your plan doesn't need to be more than a couple of pages. As an agile tech entrepreneur, I recommend your business plan include these key elements:

- Executive Summary
- Company Overview
- Business Description
- Market Analysis
- Operating Plan
- Marketing and Sales Plan
- Financial Plan

I'll describe each section below, but download the two-page business plan template from www.The60MinuteStartup.com to quickly write your plan.

Executive Summary

This comes first in business plan documents, but you write it last. It's an overview of your business, the problems you solve, your ideal customer, and your financial projections. Think of the Executive Summary as a high-level description of the company before you dive into the more detailed sections.

---

www.smallbiztrends.com/2010/06/business-plan-success-twice-as-likely.html.

Company Overview

This tells your company summary, your products and services, and your mission statement. It also talks about how you started, your position in the market, how you'll operate, and your financial goals. It's a short, succinct section that tells the reader what your business will do and how it's organized.

Business Description

This describes your company in terms of market need, your unique solution, and your value proposition. It addresses the opportunity you have with the problem you're solving and the market demand for it. Then you discuss your offer in more detail along with your pricing structure. You'll also talk about your value proposition, market need, and your solution in detail. Here's what to include:

- **Value Proposition**: If you had to sum up the value you provide to your customers, what would you say? What makes you unique? Imagine you're playing by Twitter's old rules and only have 140 characters to sum it up. Keep it to one sentence.

- **Market Need**: What problems do your customers have? How do you reduce them? If you don't know the answer to this, it'll be hard to build a business. If you don't have answers to these questions yet, talk to your potential customers. Find out what they like about your product or service. Why would they choose you over someone else?

- **Your Solution**: If someone asked you what you sell, what would your answer be? Your solution is the product or service you offer to solve the customer's problem. Describe this solution and why it's better than the alternatives.

Market Analysis

This shows how well you understand the needs of your market. It goes into the specifics of your industry and your ideal customer. You'll list things like age, gender, habits, geographic location, and buyer characteristics. You'll discuss the needs the market has, and how those needs are being met by you and your competitors.

Sometimes business owners like to do a SWOT analysis in this section. SWOT stands for strengths, weaknesses, opportunities, and threats. This analysis helps them improve their market positioning even more.

If you have direct competitors, do your research and keep your discoveries here. What products and services do your customers buy instead of yours? How is your business different from the competition? What makes your offer better than what's already out there?

Operating Plan

This section of your business plan tells how your company is organized and how you plan to develop as your company grows. It's a guide to how your business works. Who has what management responsibilities? What dates and budgets will you meet to track results? What future growth goals do you want to meet?

You'll talk about how your business will launch your product or service to the market and how you'll support your customers. This section isn't theory—it's the logistics, technology, and the grunt work involved in making your new business happen.

If you're launching a business with more than one person, you'll talk about your team in the Operating Plan section. Why did you choose certain people for certain job roles? If it's only you, write a few bullet points about why you're the right person to start and run this specific business.

Finally, if you're using partners or other resources to help your business grow, list them. Tell why they're important to your success.

Marketing & Sales Plan

Marketing and sales are two of the most important functions in any business. How will you promote your business to get sales? How will you meet your first customers? This is what you talk about in this section. Describe your key marketing messages and the channels you'll use to bring in leads for new customers. If you plan to buy ads, talk about your ad-buying strategy. Also list out your sales channels. What are the places (online or offline) where you'll sell your products? Expect this section to take shape over the next two weeks or so!

Financial Plan

This is the last section of your business plan. It's time to assign numbers to ideas. Here you list your estimated sales forecast, startup expenses, and break-even analysis. This section mentions how you'll keep the business profitable long-term and estimates your profit and loss.

As much as this section estimates revenue, it also estimates your expenses. You need to know what expenses you'll have, and how much revenue you need to become profitable. Based on these numbers, what sales goals do you need to meet to be successful? Don't worry too much about the minute details, think in broad strokes. The purpose of this section is to get a rough idea of how your business finances will work. You can refine them later.

Write Your Own Business Plan

Check off the box beside each task as you complete it.

❏ 1. Complete your business plan: 60 minutes

Download the business plan template at www.The60MinuteStatup.com. Fill out the two-page document. When you're finished, keep it in a safe place so when the time comes to make iterations you can do so intelligently.

Daily Standup

Did you complete today's tasks?

❏ Yes

❏ No

If no, what do you need to carry over to work on tomorrow?

_____

What did you learn about your business (or yourself) today that will serve you in the future?

_____

Day 10 Manage Projects to Manage Tasks

From English Major to Technology Influencer

Jill Dyché started her career as a tech writer and has worked extensively in the technology industry. Jill was a cofounder of Baseline Consulting, a management consulting firm. Baseline was acquired by SAS Institute where Jill became a Vice President. She has written three books and co-authored a fourth book, all focused on the business value of IT.

With such a technical pedigree, you might assume that Jill's education was in engineering. While that makes perfect sense, you would be wrong—Jill was an English major! How could a liberal arts major student become a tech startup founder and influencer? Well, Jill has always been a rebel. She always dared to question conventional wisdom while growing up. She did not try to figure out everything before taking off on a new adventure. She jumped in and figured it out as she went along.

For example, Jill did not put together a one hundred-page business plan before she started Baseline Consulting. She simply advised a few companies that approached of their own volition about issues they were having crafting business intelligence systems and new data strategies. Gradually, Baseline Consulting became the leader in the niche of analytics and data strategy consulting.

Today, Jill writes extensively on all things tech. One of her popular articles is entitled "Starting A Consulting Company: What I'd do differently the second time." The piece touches on five key themes.

1. Spend more time interviewing people.

2. Use more templates.

3. Speak truth to power.

4. Stop using technology for its own sake.

5. Connect clients to one another.

All these nuggets helped me grow my businesses, and I hope they'll help you as well.

Manage Projects to Manage Tasks

Running a tech company means you must juggle multiple projects, whether you're developing a software product or delivering a tech service. So it's important to figure out how you'll manage projects before you acquire customers. Fortunately, technology has advanced so much that you have plenty of free and low-cost project management options. Many of these platforms do more than just manage projects. They help you collaborate, store documents on the cloud, and schedule appointments. Here are some of my favorite project management tools.

Asana is one of the most popular cloud software project management tools that helps you schedule tasks, visualize your workflow, plan how your project pieces fit together, and quickly identify any overlaps in your schedule. Asana also allows you to upload attachments and share your projects with other people whether your colleagues or customers.

Basecamp is another popular project management app that you can access in your browser and on your phone. It gives you the tools you need to set up to-dos, a schedule, create and upload documents and files, message and chat with your colleagues, and check in regularly with your team. Basecamp can scale as you grow, helping you collaborate with your customers as well.

Wrike, an Asana and Basecamp alternative, lets you set up folders, create tasks, and assign task deadlines. It also automatically tracks how much time you spend on each task, which is really helpful if you charge by the hour or need to see how long a project is taking. Wrike lets you create and share your project schedule.

Trello helps you organize your tasks into boards, lists, and cards. You can move task cards from board to board as you complete tasks. Everyone on your team has the same access to cards, making it easy to see where a project stands and what needs to happen to move a project from inception to completion. Once you've made a card for each task you need to complete, you can give it a deadline, add an attachment, choose labels, make a checklist, write notes, and even share it with other Trello users.

Manage Your Projects Like a Pro: Now It's Your Turn

One of the best ways to get started with Project management is to start using it for your own business.

❏ **1. Research and select a project management platform: 30 minutes**

You'll not go wrong with any of the four apps I mentioned. If you need more choices, check www.The60MinuteStartup.com for more.

❏ **2. Start a project with your selected tool for your business startup: 30 minutes**

You may not have your first customer yet, and that's fine. So I suggest making your startup a project and include all the tasks associated with getting up and running. What better way to practice and master project management?

Daily Standup

Did you complete today's tasks?

- ❏ Yes

- ❏ No

If no, what do you need to carry over to work on tomorrow?

_____

What did you learn about your business (or yourself) today that will serve you in the future?

_____

Day 11 Schedule Your Time

Entrepreneur, Best-Selling Author, Digital Nomad

Gundi Gabrielle, also known as SassyZenGirl, is an award-winning Top 100 Business Author with thirteen #1 bestsellers. That is merely the tip of her achievement iceberg. Gundi ran her own two hundred-member choir and orchestra and is an acclaimed conductor and concert organist and pianist. Her business started with a simple travel blog, evolved into publishing books, and eventually transformed into an online course e-store.

How can a concert organist go on to become a tech entrepreneur? Or more importantly, why did she? In her own words, Gundi decided to become a digital nomad after a major personal crisis. She reviewed her life and decided to focus on what mattered most to her: freedom, financial independence, and world travel. It's not easy to completely switch paths after a twenty-year career in a field you love, but Gundi did just that.

It takes a tremendous amount of discipline, time management, and scheduling your time to travel the world like Gundi has and still be an accomplished tech entrepreneur. Today, we're going to take a page out of her book and implement similar best practices in your startup so you can work on what matters most.

Manage Your Time without Losing Your Mind

A challenging aspect of starting a business is managing your time. It's doubly hard if your business is a side hustle. That is precisely why I designed The 60-Minute Startup framework—so you can identify only the most important items to start your business, divide them into manageable chunks, and focus on finding and serving your initial customers.

Here are a few important tips to manage your time.

1.   Be fiercely protective of your time. Do not let anyone else manage your time. Be polite but firm when others try to monopolize your time or insert themselves into your calendar.

2. Find out the most productive time for you. My most productive time of day is the early morning. That doesn't mean it works for everyone. Find out yours.

3. Be accountable to yourself. To hold yourself accountable, join a group with similar interests. Reward yourself or a favorite charity when you accomplish your goals. Penalize yourself when you don't.

4. Design a system that works for you. Do you like to-do lists? Are you a pen and paper person or all-digital? Be honest with yourself.

5. Be aware of your "other" life. Have a frank chat with your family and loved ones about your side hustle aspirations. Make sure they're on board. Better yet, include them in your mission. Ignoring your other life is a sure recipe for disaster.

6. Make a schedule and stick to it. There is no one schedule that works for all. Make a schedule, keep it, and revisit as needed.

Here are some tools that can make your life easy without burning a hole in your pocket. These app-based solutions help you schedule meetings, monitor your time, work productively, and track your to-do list:

1. **Calendar**: Google Calendar, Schedule Once, Appointment, Calendar

2. **Time Tracking**: Rescue Me, Time Doctor, Hubstaff, Harvest, Toggl, Pocket

3. **Productivity**: Strict Workflow, Focus Booster

4. **To-Do List**: Wunderlist, Any.do

Build Your Schedule: Now It's Your Turn

Check off the box beside each task as you complete it.

❏ **1. Write down your own time management policies: 20 minutes**

Along the lines of those tips I mentioned earlier, write down your time policy. At what times throughout the day do you work best? When can you reach out to potential customers, when can you work on your business, and when can you deliver your product or service to buyers? Schedule your business around your productivity, your day job, and your life—not the other way around.

❏ **2. Research a tool that works best for you: 20 minutes**

Picking a scheduling software is not a must. If a physical notebook is better for you, so be it. But if you want, look through a few of the tools I've listed and see which one might work well for your business.

❏ **3. Prepare your initial schedule: 20 minutes**

Now get to work and schedule a few meetings using the tool you've selected. Prepare your schedule for the next few days. Pull together your to-do list.

Daily Standup

Did you complete today's tasks?

❏ Yes

❏ No

If no, what do you need to carry over to work on tomorrow?

_____

What did you learn about your business (or yourself) today that will serve you in the future?

_____

Day 12 Organize Your Passion

Small Business, Big Opportunity

Erin Shea works with Vistaprint, a leading online provider of marketing products and services to small businesses. Why am I writing about Erin in the midst of a group of passionate agile entrepreneurs? As Vistaprint's North American Marketing Director, Erin can check the pulse of thousands of entrepreneurs, some successful and others not so much.

Erin has learned that successful entrepreneurs and side hustlers can prioritize the two or three most important tasks among hundreds of to-dos. She has also seen first-hand that entrepreneurs who built a hobby into a side hustle and later into a full-fledged business are very passionate about their business and customers. But with that passion come effective organizational skills that allow them to run a profitable business.

Erin's mother built an interior design business from scratch. She remembers the troubles her mom endured, such as ordering business cards and incorporating her business before the internet was around. Erin has the advantage of translating those childhood experiences into helping other small businesses in her role at Vistaprint.

After surveying 2,000 or small business owners and side-hustlers, Erin's general startup advice is as follows:

1. Ensure your side business is something you enjoy.
2. Focus on tasks that generate revenue.
3. Build a strong social media presence.
4. Set long-term goals.
5. Leverage word-of-mouth marketing.
6. Network with people who run side businesses.[19]

What do all six tips have in common? Organization. If you're not organized, you will not enjoy your business, you will not be able to focus on money-making activities, you'll get distracted from marketing your business, you won't be

---

[19] "STUDY: Millions of Americans Have a 'Side Hustle' to Boost Their Incomes and Pursue Their Passions." Vistaprint Newsroom, March 9, 2020. www.news.vistaprint.com/side-hustle-study-us.

able to stick with goals, you'll probably drop the customer's ball and lose out on potential referrals, and you won't have time for networking. As you can see, organization is everything. So today, we're going to help you get organized,

Side Hustle or Primary Business: Organization Matters Either Way

So what to organize and how to organize? You've already completed project management on day ten and scheduling your tasks on day eleven. Today we're going to tackle the remaining to-dos.

Physical Workspace Management

Set aside a designated workspace for your business. The very act of going into that separate area will help you manage your business separately from your personal affairs and your day job. The quieter and emptier the better. Audible and physical distractions, well, distract!

Digital Workspace Management

With cloud storage so affordable, I urge you to consider Google Drive, Dropbox, or Microsoft One Drive to store your business files. If your startup is a side hustle, take special precautions to *never* access your business files from at your day job or on your employer-owned computer.

Financial and Business Management

You also need software for managing your invoices, revenue, expenses, and taxes. Your options include Quickbooks, Zoho, Xero, Wave, and Freshbooks.

Customer Relationship Management (CRM) and Support Management

As you build up your customer pipeline, you'll soon realize that you need to organize your customer information, record notes from your conversations, monitor progress towards closing the sale, and track support tickets. This is where CRM software comes in handy. Platforms popular among tech side hustlers include Hubspot CRM, Zoho CRM, Salesforce, Microsoft Dynamics 365, Freshsales CRM, and Insightly. As always, head on over to www.The60MinuteStartup.com for a complete list.

A few more organizational to-dos such as email marketing and social media will be dealt with on days thirteen and fourteen, respectively.

Let's Get Organized: Now It's Your Turn

You may not be able to complete all the tasks mentioned above in just one hour today. Still, use today to do your research and finalize the software candidates.

❏ **1. Set up your physical workspace: 15 minutes**

Identify a space in your home for your business. Make sure the area is distraction-free.

❏ **2. Get digital storage: 15 minutes**

Select your digital storage and set it up today. You'll not go wrong with any of the options mentioned earlier.

❏ **3. Set up your books: 15 minutes**

Select a bookkeeping software, sign up, and sync your business bank account, payment processor (e.g., PayPal or Stripe), and business credit and debit cards.

❏ **4. Sign up for a CRM: 15 minutes**

If you can't get to this today, that's OK. A CRM is not a must, but it will help your business scale when you're ready to grow past the side hustle stage.

Daily Standup

Did you complete today's tasks?

❏ Yes

❏ No

If no, what do you need to carry over to work on tomorrow?

_____

What did you learn about your business (or yourself) today that will serve you in the future?

_____

60

Day 13 Getting Yourself Together ... Online

$1,000 in Month One—Not a Bad Start

Back in 2015, Becky Beach was in $150,000 in debt, making $20 an hour at a stressful job, and had just had her first child. Everyone kept telling her to enjoy this time, but it seemed like life couldn't get any worse. While browsing YouTube videos one night, Becky came across an ad for a dropshipping business. Could this be the answer to her massive debt?

Becky started researching dropshipping. She already knew about graphic design and was also certified in user experience (UX) design. She took some courses and started a Shopify shop for selling handbags. Lo-and-behold, she made $1,000 her very first month. That gave her enough confidence to put more time into the business. In no time, she was making enough money to replace her income from her day job. She quit her job and got to work from home with her son.

Today, Becky is a mompreneur who coaches other moms on how to save money, start their own business, and make money online. Becky's target customers are moms who want to be at home with their kids. Most of her customers come from her email opt-ins. She'll offer something for free (like a planner) in exchange for joining her email list.

Becky's main challenge was selling, which she overcame by nurturing her email list and selling inexpensive digital products. She focused on building her email list from the get go and recommends everyone else do the same. She creates targeted email opt-ins and posts YouTube videos. One such video Becky recorded focused on her story of being in massive debt and creating her own online boutique to become debt free. Now she helps other moms start their own businesses with her blog Mom Beach.

Becky had few things going for her. She already knew how to build a decent website that attracted users. But more importantly, Becky excelled in customer service and was able to price her products in a sweet spot after a few tries. Becky is an agile entrepreneur.

Your Key Web Assets: Website, Email Marketing, Social Media

Website Design

This is where most people start sweating. They stress out over "designing a website", especially if they've never done it before. No need to panic just because Becky was a graphic design expert—you're just going to get something up and running. At this point, you don't need to worry about fancy designs. A basic website that works will do just fine. WordPress and Shopify make setting up a basic site incredibly easy, so I recommend starting there. Of course, if you really want to, it's something you can always outsource to an expert.

Email Marketing

It might not sound like much, but I can't emphasize enough how important email marketing is. Chances are, it's where your first paying customer will come from. It really doesn't matter which email marketing platform you choose, so just choose one. I've used Mailchimp in the past and I'm happy with it. Once you create an account, add a few email signup forms on your website and link them to your email marketing platform. This way, you can market to people even after they've left your website!

One Social Media (To Start)

You might not have time to tackle social media today, but think about which platform you'd like to start with. The major platforms are: Facebook, LinkedIn, Pinterest, Twitter,

YouTube, and Instagram. I have multiple chapters in the days ahead dedicated to using social media to get your first customers.

Set up Your Own Web Presence

Check off the box beside each task as you complete it.

❏ 1. Set up and "design" your website: 30 minutes

Yes, you really can do this in less than half an hour! Most web hosts will offer a "one-step" install for WordPress. After you do that, you can install a ready-made theme and edit the content on the website to reflect your business. If you don't overthink it, it'll be done before you know it.

I'd suggest spending twenty-five minutes making decisions about your website. Then, either outsource it to a web guy, or take tomorrow to develop the website fully. Think about:

- Which content management system do you want to use? WordPress or something else?
- Which theme works best for you?
- What content do you want to have published when the website goes live?

My only advice (no matter which route you choose) is don't spend much time or money on your website right now. Our focus is getting your first paying customers, and the fanciest website in the world won't guarantee that. Hard work in other areas will.

❏ 2. Integrate email marketing: 30 minutes

Every entrepreneur I've spoken to has said they wish they'd started email marketing sooner. Once your website is up and running, pick an email marketing platform like Mailchimp or AWeber, and integrate it with your website. Here are some platforms you can choose from:

- Mailchimp
- AWeber
- Constant Contact
- SendinBlue
- ConvertKit

Daily Standup

Did you complete today's tasks?

- ❏ Yes
- ❏ No

If no, what do you need to carry over to work on tomorrow?

_____

What did you learn about your business (or yourself) today that will serve you in the future?

_____

60

Day 14 Letting the World Know You Mean Business

***

Building a Community One Step at a Time

Building a community with two and a half million visits a month and a half million registered users from scratch is not easy. But that's exactly what Kunal Jain, CEO and founder of Analytics Vidhya, accomplished over a six-year period. Kunal's company is one of the largest data science communities offering courses, bootcamps, blog articles, and conferences on data analytics and data science.

The business began in April 2013 as a simple blog for Kunal to share his knowledge while heading Analytics and Data Science for a life insurance company. Over the next nine months, Kunal saw his articles generating engagement both on his blog and on LinkedIn. Across social media, other bloggers and data scientists were referencing his content. People kept coming back for these blogs and engaging with his content. He could sense that he could create a larger impact through Analytics Vidhya. That validation gave him the confidence to quit his job and turn Analytics Vidhya into a full-time business.

Analytics Vidhya grew steadily, first through online advertising and later from sponsorships for competitions and hackathons. In 2018, Kunal added courses and training as additional revenue streams. In 2019, Kunal held his first data science education conference, which attracted thousands of attendees and top companies like Intel Corp, IBM, and American Express as sponsors. At the time of this writing, the conference was one of the largest Data Science and Machine Learning conferences in India. From side hustler to global influencer, Kunal has had a dream run with Analytics Vidhya. Even though the journey hasn't always been smooth, it all started with a blog.

Launch Your Business Your Way

There is no one "right" way to launch a business. You can start with a soft launch like Kunal did or have a full-blown launch with a press release, social media blast, and even advertisements.

For almost two weeks now, you've been working methodically about building the various pieces necessary for your business to go live. But if you think about it, you may have already launched your business. You've been talking to your network about your business, getting their feedback, and validating your idea. To these people, your business already exists!

Now is the time for the broader launch. Whether you're building a side hustle or a full-blown venture, you want to tell as many people as possible about your business. Kunal did exactly that, spreading the word about Analytics Vidhya through viral blog content. Depending on your business model, you can either use email marketing, social media, or paid advertising to launch your business. Here's a quick breakdown on how to use each.

Email Marketing

If you already have an email list, you can start sending out messages about your business. If you're running promotions, let this list of people know so they can buy from you.

Social Media

It's important to be realistic about social media. You may not be able to publish on every platform every day, so pick the platform most relevant to your target audience. Kunal started on LinkedIn, but lots of people also launch on Facebook or Instagram.

Advertising

You can buy ads on Google or Facebook to get the word out about your business. If you're running a location-based business, consider ads in local publications. No matter how you get the word out, launching your business also involves engaging your prospects. You can choose a CRM (customer relationship management) software or you can use a basic spreadsheet. Track each prospect's name, where they came from, and their level of interest in what you're selling. This way you can engage with each prospect, find out how motivated they are to buy, and track your sales process with them.

Make Your Launch Happen

Check off the box beside each task as you complete it.

❏ **1. Create a graphic with a launch message: 25 minutes**

Use a free app like www.Canva.com to create a fun, easy graphic that announces your business. Write some copy to go along with this image when you share it.

❏ **2. Publish on social media: 25 minutes**

Publish this announcement image and the copy on your favorite social media platforms.

❏ 3. Create a prospect tracking list: 10 minutes

Start tracking your prospects by creating a list of interested people. Keep track of where they came from, how interested they are in your offer, and whether or not you close a deal with them. Set this up in a spreadsheet or in a free CRM software.

Daily Standup

Did you complete today's tasks?

❏ Yes

❏ No

If no, what do you need to carry over to work on tomorrow?

_____

60

Day 15 Buy a Business (Yes, It's an Option)

How to Grow a Tech Company You Didn't Start

Since 2007, Will Hankinson has done it all. He's built websites, designed video games, developed flash and mobile games, worked at two-person startups, managed Facebook game projects, taught college courses, worked at a digital agency, and invested in real estate. Will's latest business is one he didn't start himself. In 2018, Will bought www.IntroCave.com, a video intro maker popular among YouTubers. Instead of learning After Effects, any creator can use Will's automated intro maker. Just pick your colors and upload a logo, and two to three minutes later, the software renders a preview. Like it? That'll be five dollars.

Want a high-resolution version? Rendering takes about two hours, and you'll be notified via email when your high-quality intro is done—all for just ten dollars.

Why don't IntroCave.com customers just design their own intro clips? Two big barriers—tool and knowledge. Motion graphics software like AfterEffects or Cinema 4D are not easy to learn, and most industry-standard platforms require expensive annual licenses. If you need a single intro or don't have time to learn those tools, an intro maker is the perfect shortcut.

So, why did Will buy an existing business rather than start and grow a side hustle? Before he bought IntroCave.com, Will worked at a game studio and came home every night to design his own. For sanity's sake, he needed to diversify a little. He wanted to scratch a web development product itch as he's bounced back and forth between web development and game design. When the game studio folded, Will joined a previous client's digital agency. It's a boring gig, but it pays great and allows him to build up a nest egg so he can one day call it quits from regular employment, build up a small roster of products with 80 percent profit margins, and blow all that money funding weird little video games. Having a side hustle that already made money *and* scratched that itch would make Will's day job less painful.

Enter IntroCave.com. Will had spent years trying to think of the perfect "big idea" but never got there. By buying an already profitable business, Will cut out the need for motivation. He wasn't starting from zero. He already builds products from scratch at his agency job, so he wanted a project he could optimize and grow. Doing so hasn't been as easy as Will hoped, but still, it's been worth it.

Will expected to take over IntroCave.com and have fun adding new intro templates and writing blog articles to boost traffic. But then a Google algorithm update a few months in tanked the website's organic traffic. Overnight, revenue collapsed from $4,000 to $5,000 per month down to $1,000 per month. At the same time, most marketing emails to IntroCave.com ended up in spam folders. Apparently you can't just blast out 40,000 to 50,000 emails once a month! So Will put all his other fun side projects on hold and treats IntroCave.com like a second job. He's spent the last year learning search engine optimization tactics, building his own marketing automation and newsletter tools, and trimming expenses such as the previous marketing email platform.

If Will had to do it all over again, he would have structured the earn-outs differently for the purchase payments of IntroCave.com. He didn't have a sense of how volatile search traffic could be. Since organic search was and still is the primary way customers find the business, Will wishes he'd done his due diligence researching the threats to the particular website he was planning to buy. Nevertheless, IntroCave.com is working out, revenue is improving, and Will is playing the long game to win.

Buy Your Way into Tech Entrepreneurship

The fastest way to become an entrepreneur is to buy an existing business like Will Hankinson did. Especially if you want to build an empire while you still have a day job. In *The 60-Minute Startup*, I told the story of my first online business, www.ChoicePetMeds.com, which I actually bought to get my entrepreneurial feet wet. I grew it, sold it, and used the proceeds to fund another venture.

You're probably wondering why I'm including this whole "buy a business to become an entrepreneur" in this book. The fact is, buying an existing business is a no-brainer idea if you have money to spare but no time to start one from scratch. Like Will, maybe you don't want to wait a whole month to see your first revenue, and you'd rather start your entrepreneurial journey much further down the road. I applaud you. And I'm not being cheeky.

So the next question is, where can you buy a business? I recommend established marketplaces to purchase a domain where you can build your business, a relatively new startup, or even a full-fledged tech business with a strong sales track record. Whichever marketplace you choose, remember the age-old adage *caveat emptor.* Let the buyer beware. Do your due diligence if you want to go this route.

Best Places to Buy an Online Business

- www.Exchange.Shopify.com
- www.Flippa.com
- www.BizBuySell.com/Internet-Companies-for-Sale
- www.WebsiteBroker.com
- www.FreeMarket.com/Sites

Buying a business through online brokers is another option. The advantage of working with a broker is that it's their job to pre-vet business opportunities for you and guide you through the volatility and add clawbacks or incentives into the deal purchase agreement so you don't repeat Will's story and find unpleasant surprises in the days after you close.

Best Places to Find Business Brokers

- www.FEInternational.com (Will used this site to buy www.IntroCave.com.)
- www.EmpireFlippers.com
- www.DigitalExits.com
- www.SideProjectors.com

If you think you might like to buy an online tech business, look specifically for businesses with multiple streams of income, not just one. For example, my pet medicine business sold prescriptions but also pet food and toys for dogs, cats, and other common household pets. You want diverse traffic sources in addition to products and services. For example,

Will has turned to social media advertising to help get his traffic numbers back where they were before Google's search algorithm update. I also recommend you look for consistent revenue over at least twelve months. If a business you're thinking about buying made $50,000 this month but only $1,000 last month and fifty bucks the previous month, buyer beware!

If you do plan to go this route, use the remaining days of this thirty-day journey to get possession of the business and learn all aspects of revenue, traffic, growth opportunities, and potential threats. The remaining chapters will still be of great help to you because you, like entrepreneurs starting from scratch, want to get more customers and grow the business.

Buy an Existing Business: Now It's Your Turn

To be fair, it will probably take more than one hour to do the proper due diligence before you buy a business. By now, you should know what kind of tech business you may be interested in. Use today to review businesses for sale. Here's how to identify the best deal:

❑ 1. Research businesses for sale: 60 minutes

Dig into the business marketplaces and brokerage websites mentioned in this chapter. Look through businesses for sale in your domain of expertise. As you find businesses in your budget, do your due diligence. Either research the growth opportunities and potential threats or hire a broker to do so for you.

Daily Standup

Did you complete today's tasks?

❏ Yes

❏ No

If no, what do you need to carry over to work on tomorrow?

_____

What did you learn about your business (or yourself) today that will serve you in the future?

_____

Day 16 Customer Development

If You Build It, They Will Come (No, They Don't)

In the movie *Field of Dreams*, Iowa farmer Ray Kinsella (Kevin Costner), mows down a large part of his cornfield to build a baseball diamond. His dream was that all the baseball legends will come and play once he builds the baseball park. Of course, it took more than building the park for them to come and play.

Many technology startup founders, myself included, have launched businesses with an illusion that once we build this ground-breaking product, the customers will come in droves. That's just not how customers buy. Steve Blank, author of *The Four Steps of the Epiphany*, explains customer development in four steps:

1. Customer Discovery
2. Customer Validation
3. Customer Creation
4. Company Building

In this book, we have been focusing on customers since day one.

Sukhi Jutla from day two walked around the Diamond market in London asking her prospects how they buy their diamonds and their pain points even before she had any product or service. Only when she validated her solution did Sukhi start working on her company.

On day dive, you met Vartika Manasvi, who had a bare-bones website as a proof of concept to pitch to her first prospect who then converted into a paying customer even before she had a full-blown product. Both Sukhi and Vartika followed the Customer Development methodology from the outset.

Customer Development: A How-To Guide

The very first step in the process of customer development is customer discovery, which can be divided into two parts.

1. Finding ideal customers (who and where)
2. Interviewing them to convert them into paying customers (what and how)

In this chapter, I'll focus on part one. Tomorrow is the second part.

Here is a comprehensive list on where you can find your initial customers. We'll be going through each one of these in more detail in the coming chapters.

- **LinkedIn**: Ideal for B2B customers. Join LinkedIn groups. Use LinkedIn Answers. Work with your connections. Run LinkedIn ads.

- **Facebook**: Ask your friends. Join relevant Facebook groups. Research fan pages. Run Facebook ads.

- **Twitter**: Ask your followers. Search for relevant hashtags. Join Twitter chats. Run Twitter Ads.

- **YouTube**: Search for channels in your industry. Look at the comments. Research commenters' channels. Direct message any who look like ideal customers.

- **Email Newsletter**: Use your signature to ask questions. Start a personal newsletter.

- **Meetup.com**: Join & Attend meetups in your category. Request organizers to introduce you and send messages on your behalf. Message users on Meetup.com. Create a meetup group.

- **Blog**: Start and write a blog post. Share your blog posts on other platforms like Linkedin and Reddit. Write guest blog posts.

- **Q&A Websites**: Use Quora, Quibb like sites and reach out to people who ask relevant questions. Answer questions in your niche.

- **Craigslist**: Select categories that your business operates in. Reach out to posters. Make your own post.

- **Forums, Micronetworks, and Communities**: Join relevant forums for your business. Post questions or answer questions. Participate in discussions.

- **Google AdWords**: Run AdWords with a simple landing page or relevant keywords.

- **Startup Accelerators**: Join relevant accelerators like Y-Combinator and connect with other startups.

- **Your Competition**: Study your competition. Look for social mentions on your competitors.

- **College, University, School**: Connect with your friends and classmates. Ask them to spread the word around.

- **Kickstarter**: Look for similar products being funded. Reach out to users who funded them. Start your own funding campaign.

Start Your Own Customer Discovery Journey: Now It's Your Turn

Check off the box beside each task as you complete it.

❏ **1. Choose three platforms from the list to do your research: 60 minutes**

The above list is a fairly comprehensive and exhaustive list. Your goal is to focus on a select few. As this is a very important step and the remaining chapters will go into specific details on many of these areas, take your time today to do thorough research.

Daily Standup

Did you complete today's tasks?

❏ Yes

❏ No

If no, what do you need to carry over to work on tomorrow?

_____

What did you learn about your business (or yourself) today that will serve you in the future?

_____

Day 17 Customer Discovery

Do You Really Know What Your Customers Need?

You started on your customer journey on day four asking simple questions like, "Who is your customer?," "What are their pain points?," and "How can your product or service solve their pain points?" Many businesses focus on what customers want and try to address them. Only the astute business owner knows that a customer's "want" is not necessarily the same as their "need." By asking relevant customer discovery questions the right way, you can build a massive customer pipeline.

John Roberts and William of timesheet company WeWorked launched their company assuming they would attract mostly US-based customers. Then they noticed that people from Europe kept signing up. Curious, they asked them simple questions such as, "What do they like about the software?"

Nathan Miller of Rentec made his software freely available to other landlords. After he got an inquiry from a property manager, Nathan wanted to know why property managers would be interested. Soon he realized property managers had nothing like his software and desperately needed it.

How to Ask the Right Discovery Questions (The Right Way)

In the last chapter, you shortlisted digital and offline places you can find your ideal customers. Now what? Engage them in conversation to discover their true pain points and needs. If you pester and nag with dumb questions, you'll quickly turn them away. For example, many marketers on LinkedIn ask me if I'm interested in their product brief or a video right after connecting with me. Really? I'm turned off right away and never read their messages after the pitch.

Depending on what you know about your target customer's needs, use the following template to assess which questions you should ask your prospects.

Customer Identification

Make sure the customer is within your segment. Get to know more about their demographic. A couple of example questions are:

- What do you do?
- Who handles (*your service or product*) at your home or office?

Discovering Problems

These questions validate your hypothesis about a problem or will help you learn about problems. Example questions include:

- What product do you wish you had that would help you do your job better?
- What tasks take up the most time in your day?

Problem Validation

If you don't yet know the customer's problems, ask questions to find out. Example questions include:

- Do you find it hard to (*process or problem*)?
- Tell me about the last time you (*tried to solve the problem*)?

Product or Service Validation

In this section, you'll validate your assumptions about your product or service. Example questions include:

- What do you think about this product?
- Would you use this product?

Here are some tools you can use for your customer discovery.

- **Google Forms**: Google has created a simple, easy, and free tool to put together a questionnaire.

- **Survey Monkey**: Create and deploy surveys using this popular tool.

- **www.UserTesting.com**: Once you have a minimum Viable Product (MVP), get feedback from real people.

- **www.Appsee.com**: If you're working on a mobile app, this platform will help you collect data on how users interact with it.

For a complete list of customer discovery questions and tools, head on over to www.The60MinuteStartup.com.

Discovery: Now It's Your Turn

Check off the box beside each task as you complete it.

❑ **1. Prepare your customer discovery questions: 30 minutes**

First ask yourself how much you know about your prospects. Choose a question section based on your answer. For example, if you don't know much about your customer, ask customer identification questions. If you are unsure about their problems, focus on the problem identification section. And so on.

❏ **2. Select one or two tools to deploy your questions: 30 minutes**

If you already have an email list or contacts, you can start with Google Forms. If you need access to a curated list, you can rely on platforms like AppSee, which will give you access to real customers for a fee.

Daily Standup

Did you complete today's tasks?

❏ Yes

❏ No

If no, what do you need to carry over to work on tomorrow?

_____

What did you learn about your business (or yourself) today that will serve you in the future?

_____

60

Day 18 The Most Likely First Customer

Family, Friends, and Followers: Your Most Trusted Network

The hardest part is over. You've launched. You now have a business. Time to breathe a little easier. Now on to acquiring customers. Over the next several days, you'll learn several strategies to find your initial customers. You may not need every strategy to book yourself solid for the next several months, however. I recommend you focus on the three to five strategies that you're comfortable with and are most likely to work for your type of business.

On day four, you started to think about your prospective customers and how your product or service could address their pain points. I advised you to compile a list of people from your immediate connections such as family, friends, and colleagues. These are your likeliest first customers—even if they don't match your ideal customer profile. Why do I think so? Because I know from first-hand experience.

One of my first side businesses was website design. To get the word out, I told everyone I knew about the website I'd built for myself as proof I could do the same for others. Soon after, the tenant from my investment property reached back out to me. Unbeknownst to me, she was planning to start her own fundraising consulting business for Catholic charities, and she needed a website. Bingo, I had my first customer.

Numerous guests on my Agile Entrepreneur podcast converted family, friends, and colleagues into their initial customers. Nicole Landau of Landau Consulting Solutions received a referral from a colleague, who turned into her first accounting client. Jason Patel's initial customers for his academic tutoring service were friends' referrals.

There are several reasons why your immediate network is the best place to find your first customers. First of all, they trust you. Family and friends know you and like you. They believe in you and want you to succeed. Second, they probably already know people who are looking for a product or service like yours. According to a Harris Poll, 82 percent of people ask people they know for recommendations before they buy *anything.*[20] And third, if people believe they can benefit from your great offer, they're going to buy! Simple yet true.

How to (Or Not to) Pitch People You Know

If you're like me, you're not exactly comfortable about the idea of pitching your business to people you know, whether it's your boss, your gold buddies, or your own mother. My advice? *Don't pitch.* Just inform. Tell your closest connection what you're up to. Your new business, your product or service, and who you're out to help with it. That's all. Take this honest, straightforward path, and you'll realize that you don't have to beg for business right off the bat (that's a definite no-no anyway). To let your network know about your business without turning them off, I recommend the following four steps:

1. Call or set up a meeting over coffee, lunch, or a home barbecue. At this casual meeting, simply tell family and friends about your business, your offer, and why you're doing this in the first place. The primary purpose is to build up excitement and create interest. If you sense an opening for feedback, ask for advice.

2. To people you can't meet in person, send a personalized email or letter explaining your business plans and request honest feedback.

---

[20] Kapadia, Amity. "Numbers Don't Lie: What a 2016 Nielsen Study Revealed About Referrals." Business 2 Community, March 12, 2016. www.business2community.com/marketing/numbers-dont-lie-2016-nielsen-study-revealed-referrals-01477256#XHrEtgC6lyrOB397.97.

3. As your business makes progress every day, keep your network informed. I mean weekly, not daily. Nobody likes spam. Keep it brief but focus on the highlights of your journey. It's OK to mention the downsides; everyone knows that entrepreneurship is not easy and has its ups and downs.

4. If possible, tell family and friends about your discount offers, early bird specials, or referral incentives. But don't pitch directly unless they ask. You want to generate interest, not push your business in anyone's face. The goal is to stir up excitement so the people you know refer potential customers that you don't, giving your business a much larger reach.

Leveraging Your Trusted Network: Now It's Your Turn

You know who these people are. Now it's time to put them to work for you.

❏ 1. Update your friends, family, colleagues list: 15 minutes

Pick up with the list of family, friends, and colleagues you prepared on day four. Update the list by asking these questions:

❏ Who are your personal friends?

❏ How about any classmates? People that you came across in schools and colleges?

❏ Your business connections? Colleagues? (Avoid your employer's customers because of a conflict of interest.)

❏ Contacts within clubs or religious institutions you're affiliated with?

❏ People you run into on a daily basis? Your lawyer, hair stylist, doctor, etc.?

❏ Neighbors, both past and present?

❏ People from your hobby groups? Hiking, running, bicycling groups?

❏ **2. Set up meetings or craft your personal message: 30 minutes**

Call or email your list. Either set up meetings or simply send messages. Remember, your goal is to convey your excitement to your most trusted contacts.

❏ **3. Update the status of your calls and emails: 15 minutes**

Use this time to update the status of your outreach and, if necessary, refresh your list with additional contacts.

Daily Standup

Did you complete today's tasks?

- ❏ Yes

- ❏ No

If no, what do you need to carry over to work on tomorrow?

_____

What did you learn about your business (or yourself) today that will serve you in the future?

_____

Day 19 Alpha and Beta Testing to Find Customers

Test Before Your Sell

"**D**on't judge a book by its cover."

Nobody follows this rule. When you browse books online or in stores, you click or pick up the book with the most attractive covers. Knowing this, I asked my graphic designer to mock up several versions of *The 60-Minute Startup* before publication. We used different color schemes and layouts for these covers, and I asked several aspiring entrepreneurs for feedback. This is called "beta testing." It's used extensively in software and hardware development. Why beta? Well, depending on which stage of product development you decide to test, they're either alpha tests (early stage) or beta tests (later stage). Beta testing has two primary goals:

1. To get early feedback before the product or service is released.

2. To get prospective customers to try the product before they purchase.

Nathan Miller, the CEO and founder of property management software provider Rentec Direct whom you met on day four, beta-tested his application. He offered platform access to several landlords he knew, his ideal clients. They liked the application so much that they became paying customers when Nathan ended testing. As a bonus, these landlords became Rentec's biggest advocates, telling other landlords to sign up for this great new application.

Another beta testing success story is Pinterest, the image-sharing social media app. The founders listed the early stage version on a testing site called Betalist, beta testers loved it, and the rest is history.

The Nitty-Gritties of Beta Testing

So, how do you beta-test your offer? Earlier in your journey, you listed family, friends, and colleagues. The cheapest, easiest, and fastest beta test you can run is to ask for their feedback. If people in your network don't match your ideal customer profile, you can use a beta test website. Here are some of the most popular:

- **www.Betalist.com**: Betalist is the granddaddy of all startup beta sites. Your product or service must meet their selection criteria, and they charge a fee to feature your business.

- **www.ProductHunt.com**: Product Hunt surfaces the best new products every day. It's a place for digital product enthusiasts to geek out about the latest mobile apps, websites, hardware projects, and tech creations. The platform combines peer referrals with a forum to highlight intriguing new products available for beta testing.

- **www.AlternativeTo.net**: AlternativeTo lets users find new Windows, Mac, Linux, and online software applications via crowdsourced recommendations. If you're building software or an app, AlternativeTo might be a great place to beta test.

- **www.BetaPage.co**: BetaPage is a community where you can list your startup to get feedback from a huge audience of founders, entrepreneurs, investors, incubators, mentors, indie hackers, and freelancers.

For a comprehensive list of beta test websites, visit www.The60MinuteStartup.com.

A beta test that converts users into customers follows these four steps:

1. Identify your ideal target customer. Be precise about the demographics (age, sex, education, income, profession, etc.) and psychographics (dreams, goals, values, etc.).

2. Make your beta test group exclusive so members feel special.

3. Offer a discount or early bird promotion if beta testers sign up once the test is complete.

4. Create some buzz! Publish the beta test results on social media and share them with your trusted network.

Beta Test Your Way to Your First Customers: Now It's Your Turn

❏ 1. Identify your beta test target audience: 15 minutes

Do you want to use your immediate network for your beta test, or do you want to use a platform like Betalist?

❏ 2. Set up your beta test environment: 45 minutes

Devote the remaining time today to setting up your beta test. If you're planning to use a beta testing platform, pick the one best suited for your offer, register your tech startup, and fill in all the information they ask for. If you plan to reach out to your network instead, think about how they will beta test your offering and set up that environment. For example, with *The 60-Minute Startup* cover options, I labeled each cover with a number, attached them to an email, and asked people which cover they preferred. I added up results and settled on my final cover based on beta testers' feedback.

Daily Standup

Did you complete today's tasks?

❏ Yes

❏ No

If no, what do you need to carry over to work on tomorrow?

_____

What did you learn about your business (or yourself) today that will serve you in the future?

_____

Day 20 The Fastest Way to Your Customers

Is Paid Advertising a Scam or the Best Thing Ever?

If you search online for tips to get customers, you're going to find every conflicting opinion on paid advertising. Some people claim they're a total waste of money. Others say online ads are effective to get only your initial customers. Still, some say the opposite, that paid advertising only works for established companies looking to grow. I see glimmers of truth in each opinion. Personally, I believe that paid advertising is **best** for technology consulting, technology services, and technology training businesses. Here's why.

Organic social media reach is down.[21] That means fewer people see what you post. Hence, fewer and fewer returns on your time investment. Only paid advertisements get fast results. When you advertise, your money guarantees reach. And with digital ads, you don't have to guess if your message is hit or miss. Social media platforms like Facebook's Business Manager allows you to measure every ad's performance so you can fine-tune your ads to boost return-on-investment while lowering your ad spending. You can build your first ad in minutes, budget as low as five dollars per day, and track your results from there. Whether you use Business Manager or another online advertising platform like Google AdWords, you get access to sophisticated targeting tools that allow you to spend your budget reach a specific audience of ideal customers only.

In *The 60-Minute Startup*, I profiled digital nomad Ali Saif, founder of www.HighClickz.com. Ali's agency runs profitable Facebook and Google advertisements for clients just as he uses them to grow his own business. To see how Ali practices what he preaches and learn the secrets of paid advertising success, check out day seventeen of *The 60-Minute Startup*.

Another successful advertiser you can learn from is Brian Meert, founder of Advertisemint.com. You know Brian

---

[21] Bernazzani, Sophia. "The Decline of Organic Facebook Reach & How to Adjust to the Algorithm." HubSpot Blog. Accessed April 7, 2020. www.blog.hubspot.com/marketing/facebook-organic-reach-declining.

from day ten of this book. He built his own company using Google AdWords. As his advertisements converted paid search traffic to his website into paying customers, he added Facebook ads as well. Now, Advertisemint offers paid advertising services based on the exact strategies Brian used to grow his company. There is nothing quite like customers on-demand, and only paid advertising offers it.

Which Advertising Channel is Best For You?

Almost all social media platforms provide paid advertising options. So which one is the best for your tech startup? It probably comes as no surprise that Facebook and Google dominate, with 50 percent of market share between them.[22] In my experience, the three most effective platforms to get your initial tech business customers are:

1. **Facebook Advertising**: Ideal for entrepreneurs who know who their target customer is, as Facebook's Business Manager platform lets you target those precise customers.

2. **Google Advertising**: Very effective for local businesses, high-search volume products and services, and trending markets to reach customers who are actively looking for solutions to their pain points.

---

[22] Liberto, Daniel. "Facebook, Google Digital Ad Market Share Drops as Amazon Climbs." Investopedia, June 25, 2019. www.investopedia.com/news/facebook-google-digital-ad-market-share-drops-amazon-climbs.

3. **LinkedIn Advertising**: Great for business-to-business (B2B) outreach and for connecting with a professional audience.

Here is a simple cheat-sheet for figuring out which platform is best for you.

Advertising Platform	Audience Profile
Facebook Advertising	More than 2.45 billion active users making personal connections with family and friends[23]
Google AdWords	Very broad audience mostly searching for information before buying
LinkedIn Advertising	Professional audience looking for career, education, and connections

Advertise Your Way to Your First Customers: Now It's Your Turn

❑ 1. Decide on your preferred advertising platform: 15 minutes

As you decide on your preferred platform, answer two key questions: Who is your target customer? And where do they spend most of their time on social media? Refer back to the Audience Profile column to help.

---

[23] Sehl, Katie. "27 Facebook Demographics That Matter to Marketers in 2020." Hootsuite Social Media Management, December 20, 2019. www.blog.hootsuite.com/facebook-demographics.

❏ 2. Set up your first ad: 45 minutes

Each platform has its own ways to set up ads. This book's companion website www.The60MinuteStartup.com has tutorials for Facebook, Google, and LinkedIn advertising. Select three to five marketing and sales strategies. Paid advertising should probably be one of them. If this particular strategy takes more than forty-five minutes, that's OK. The basic idea is to test, tweak, and iterate. That's the agile way.

Daily Standup

Did you complete today's tasks?

❏ Yes

❏ No

If no, what do you need to carry over to work on tomorrow?

_____

What did you learn about your business (or yourself) today that will serve you in the future?

_____

60

Day 21 Find Your Tribe

Internet Tribes: The New Normal

The word "tribe" probably conjures up the image of a small band of closely related people hunting, gathering, and working the land to survive. Tribal communities still exist in some parts of the world today, but the vast majority of us are enthusiastic members of an entirely different kind of tribe—a digital tribe.

Social media specifically and the internet generally have given us the ability to form unofficial, loosely affiliated communities with other people who share interests. Today, I want to single out two platforms that have become popular places to start and grow technology-focused internet tribes, Quora and Reddit.

Quora is a question-and-answer website where anyone can ask, answer, and edit questions on any topic. Every month, over 35 percent of the US population and approximately 300 million users join in these spontaneous discussions, which regularly turn into essay-length back-and-forth conversations between the original question poster and one or more responders.[24] The number one most followed topic on Quora is Technology.

Better known than Quora is Reddit, the sixth most popular website in the US.[25] Reddit bills itself the front page of the internet. The site is a massive collection of forums where users can share links, photos, and videos and comment and vote on other people's content.

So why should a tech side hustler like you pay attention to Quora and Reddit? You have an advantage on both platforms. The tribes these sites attract tend to be more tech-savvy than users on other social media. That's why they can be a gold mine for your business—if you approach each platform the right way.

---

[24] Archibald, Maggie. "15 Quora Statistics For Marketers In 2020." Foundation Marketing. Accessed April 7, 2020. www.foundationinc.co/lab/quora-statistics/.

[25] Widman, Jake. "What Is Reddit?" Digital Trends. Digital Trends, March 11, 2020. www.digitaltrends.com/web/what-is-reddit/.

Using Quora and Reddit to Get Customers

Personally, I like Quora more than Reddit for two reasons. Number one, Quora is a neutral platform strictly focused on asking the right questions and answering them with data, facts, and details. And number two, Reddit is a platform of opinions. Snarky content tends to perform the best. Anonymous users with attitude can bury a beginner if they don't like what you post about your business.

Here are four steps to acquire customers on Quora:

1. Complete your profile on Quora with details about your business and contact information.

2. Build your expertise. Search for questions related to your business, find unanswered questions, and answer with useful details. Don't self-promote. Let your answers and the ensuing engagement speak for itself.

3. Ask thoughtful questions yourself. The beauty of Quora is that you can get a lot of responses with insightful questions. For example, you could ask users about their experiences with problems that your business solves. When people answer, keep the conversation going. Either reply to their answers or go to the next level—visit their profile and send a direct message.

4. Keep it up. The more you show up, the more likely you'll get noticed and, therefore, the more likely it is you'll find customers. Or in the case of Quora, they find you!

Five steps to use Reddit to find paying customers:

1. Create a Reddit profile. Include information about your business and how best to learn more about what you do.

2. Find "subreddits" (niche communities) related to your business. Relevant subreddits for your tech startup might be r/entrepreneur, r/technology, and r/smallbusiness.

3. Curate viral content. If you find creative, funny memes or gifs that your prospects might relate to, share them. On Reddit as on other platforms, quality content can go viral quickly.

4. Build your Karma—Reddit's term for your credibility. You get Karma for the upvotes your content receives. You also earn Karma when you comment on other users' posts. More engagement, more Karma.

5. Promote deals, discounts, and giveaways on the subreddit r/deals.

Build Your Own Internet Tribe: Now It's Your Turn

❏ 1. Complete your profiles: 15 minutes

You can choose to focus on both Quora and Reddit, but I'd recommend starting with the one that seems to suit you better. Only you can make that call, but there is no wrong answer.

❏ **2. Research topics of interest: 15 minutes**

Look for unanswered Quora questions and/or relevant subreddits. On Quora, read related questions to see if there's a common theme. What do people most often want help with? On Reddit, familiarize yourself with the most popular content on the question to get plenty of ideas for your own posts.

❏ **3. Get building: 30 minutes**

Invest half an hour today preparing your posts, answers, and other forms of engagement. Publish your first post. Voilà! You are now on your way to getting your initial customers. Quora and Reddit both reward consistency, so if your first few posts, answers, or comments get traction, you know you should keep it up.

Daily Standup

Did you complete today's tasks?

❏ Yes

❏ No

If no, what do you need to carry over to work on tomorrow?

_____

What did you learn about your business (or yourself) today that will serve you in the future?

_____

Day 22 Search Engine Optimization for the Win

I'm sure you've heard about search engine optimization, SEO for short. But if you're like most people—even many entrepreneurs—you may not know exactly how SEO strategies work to boost online content to those coveted first few search result spots. First things first. The *what* of SEO is how high your website or web page shows up in unpaid search results. The *how* is optimizing that site or page so it performs well in organic search.

In my experience, SEO takes time to yield results. Because search engines like Google regularly update their algorithms without much notice, SEO is an imprecise, ever-changing science. So why are we talking about SEO as a great way to get customers by day thirty? The fact is, if you want to generate leads via your website or attract primarily local customers, SEO is beneficial. And if you're in the tech startup, professional service, online business, local electronic repair and service, or Software as a Service (SaaS) space, smart SEO is essential. Done right, SEO can generate leads in a short period of time. Yes, even in a matter of weeks, sometimes days.

When it comes to getting your business noticed, SEO is the biggest game in town, and Google is its superstar. Of all searches, Google accounts for 78 percent. And it averages 100 billion searches a month with 3.5 billion every day.[26,27] SEO beats other ways to acquire customers. Leads who find a business through an internet search become customers 14.6 percent of the time; compare that close rate to outbound lead generation tactics like paid advertising, direct mail, or phone calls, which average a paltry 1.7 percent close rate.[28] SEO is just the smart way to grow a business!

For example, Will Hankinson from day fifteen relied on SEO to grow the business he purchased, Introcave.com. If you

---

[26] "How Many Google Searches Per Day?" Serpwatch.io, March 4, 2020. www.serpwatch.io/blog/how-many-google-searches-per-day/.

[27] Bond, Conor. "27 Google Search Statistics You Should Know in 2019 (Insights!)." WordStream, February 26, 2020. www.wordstream.com/blog/ws/2019/02/07/google-search-statistics.

[28] "A Practical Guide to SEO Metrics." Digital Marketing Institute, November 15, 2019. www.digitalmarketinginstitute.com/en-us/blog/practical-guide-seo-metrics-2.

are thinking about buying an existing business, you need to know SEO to acquire new customers. You want that investment to pay off quickly without having to dump more of your own money into the venture through paid advertising. And remember Becky Beach from day thirteen? She taught herself SEO to quickly build her online business to $1,000 per month in just a few months. If these entrepreneurs can convert search traffic into paying customers, so can you. Here's how.

Pick the Low-Hanging Search Fruit

SEO broadly consists of two categories, on-page optimization and off-page optimization. You have more control over on-page optimization because you decide what is and is not on your website or web page. You can also make optimization tweaks quickly. Anyone can learn these on-the-fly on-age search optimization strategies:

1. **Write unique high-quality content**. By far, this is the most important factor. Make sure that you have high quality content on your website that is unique. Also make sure that the content includes some relevant keywords that you want to target for your business. Keyword research is an important topic but beyond the scope of this book.

2. **Decrease page load time**. Google prefers web pages that have a rapid page load speed. Within fractions of a second of a user clicking on a link to your website, the page loads. So how do you find out your page load times, and how do you improve them?

Google itself has your answers. Check out their free page load time tool Pagespeed Insights to learn how quickly any page on your website loads. Pagespeed Insights also gives your page a rank and suggests ideas for improving user experience on your website. It's basically a free website audit from the world's smartest internet company!

3. **Make your website mobile-friendly**. Over 52 percent of all web browsing is done on smartphones and other mobile devices.[29] How does your website look on a phone? On a tablet? Luckily, Google offers a fast and free Mobile-Friendly Test to judge the mobile version of your website and specific action steps you can take to improve user experience.

4. **Use on-page SEO optimization**. Free and low-cost tools like Yoast SEO enable you to quickly scan your website for broken links, missing keywords, and other quality issues that otherwise warn Google not to rank your website near the top of search results.

Off-page optimization is what makes SEO a long game. You have less control over what other websites do to impact your search results ranking, *but* there are activities you can do that will, over time, get Google and other search engines to notice your website and, let's hope, boost your website in the results. Key off-page optimizations include:

---

[29] Clement, J. "Mobile Share of Website Visits Worldwide 2018." Statista, July 22, 2019. www.statista.com/statistics/241462/global-mobile-phone-website-traffic-share/.

1. **Earn backlinks**. Backlinks are links to your web pages and posts from other websites. The quality of the websites where your content is linked influences how search engines rank your website. I recommend identifying relevant websites, guest posting there, and getting backlinks (e.g., your website in your article author bio). It does take time to execute though because you don't have full say in whether or not another website publishes your content. My advice is to do what you can during this first month of your business to build a few backlinks on quality sites that already get lots of traffic.

2. **Show up on social media**. Social media is a totally different topic by itself, but in the context of SEO, make your website content social media-friendly so people can easily share it. For example, an infographic is 30 percent more likely to be shared than a post without one.[30] Likewise, any visual content including video is much more likely to be shared. If you want your website to have a blog, great! Tools like Social Report, Buffer, MeetEdgar, HootSuite, and CoSchedule help you schedule posts to be shared on social media.

---

[30] Carlson, Amanda. "32 Stats & Facts That Prove Infographics Aren't Dead." Lucidpress Blog, June 21, 2017. www.lucidpress.com/blog/32-infographic-stats-facts.

Search Engine Optimization: Now It's Your Turn

Yes, SEO is a daunting task. Let's not pretend sixty is enough time to apply what you've learned in this chapter and get to page one. But what you can do with your hour today is set yourself up for SEO success. Get your business ready for longer-term search success, and you'll have everything in place you need to execute over the next few days.

❏ 1. On-page optimization, content ideas, page speed, et. al: 30 minutes

Use this half hour to generate content ideas you can write about over the next few days. Also check your website page speed and mobile-friendliness to see if you easily and quickly boost your business' visibility.

❏ 2. Keyword and backlink research: 30 minutes

Finding out which keywords to use on your website and looking for quality websites where you can guest post takes more than thirty minutes. Again, that's the minimum time you should invest in this task today. But if all you do is find *one* high search volume keyword and work it into your website's home page and pitch *one* industry website an article idea, you'll be much further along than anyone who takes one look at SEO, thinks it's too hard, and gives up. And in a few months' time, your business will be getting noticed by prospective customers while they're having to think about shutting down.

Daily Standup

Did you complete today's tasks?

❏ Yes

❏ No

If no, what do you need to carry over to work on tomorrow?

_____

What did you learn about your business (or yourself) today that will serve you in the future?

_____

Day 23 The Right Social Media Platform for Your Business

You Can Run But You Can't Hide from Social Media

Up until 2016, I did not have Facebook or Twitter accounts. I'd uploaded only two videos to YouTube, and those were my daughters' school performances. I had fewer than 500 connections on LinkedIn. Somehow, businesses were finding me. I received unsolicited marketing emails and phone calls to buy this or that from businesses I'd never heard of. How had they gotten my contact information?

One day, I googled my name. Holy cow . . . what a surprise! The search results showed where I lived, where I went to college, where I worked, and so on. Apparently, old friends and former classmates had been posting about me on social media, including pictures to Facebook. I realized I could run but not hide from social media. That same day, I decided to embrace social media.

Fast forward to today, and social media is an essential part of my life. As a business owner, I use social media every day to engage existing customers, comb for prospects, and deepen new relationships. My two favorite platforms for groups are Facebook and LinkedIn, where I run my own private communities. Of course, I wouldn't continue to use social media if my efforts weren't paying off. The same should go for you.

In *The 60-Minute Startup*, I told the story of Robyn Mancell, founder of an online foreign currency exchange academy called Girls Gone ForEx. She used free Facebook contests to give away her very first course memberships. Contestants were so excited about the academy that several people who did not win actually asked to pay to join! On day thirteen of this book, you met Becky Beach, who used YouTube and Pinterest to build her online ecommerce business. Social media is full of customer acquisition opportunities as long as you use the right platform the right way.

Which Social Media Platform Is Best for Your Tech Startup?

Facebook, Twitter, YouTube, LinkedIn, Pinterest—the day has only just begun, and already you have five social media platforms to choose from. And these five are a few websites among many where you can find new customers. If you feel overwhelmed, I can understand. With so many choices, it's hard to decide which one is right to build your presence.

To help you get started, I prepared a cheat-sheet that focuses on the top six social media platforms. Personally, I consider Facebook, Twitter, and LinkedIn to be the best platforms for technology entrepreneurs to see return on investment. For a more exhaustive list of social media platforms, head on over to www.The60MinuteStartup.com.

Facebook

Total # of users: > 2 billion users

Demographics: Ages 25-54; 60% female

What users do: Build relationships and keep in touch with family & friends

Type of content: pictures, videos, contests, short posts

Good for: Building a dedicated following, building brand loyalty

Twitter

Total # of users: > 330 million users

Demographics: 18-49 years

What users do: follow trending topics using hashtags

Type of content: news, short conversations

Good for: Brand awareness, market research, public relations

YouTube

Total # of users: > 2 billion users

Demographics: All ages

What users do: Learning,

Type of content: How to videos

Good for: Brand awareness, service industry, entertainment

LinkedIn

Total # of users: > 600 million users

Demographics: 30-49 years

What users do: Build relationships and keep in touch with family & friends

Type of content: long articles as well as posts, news, professional conversations

Good for: Business development, B2B lead generation, career services

Pinterest

Total # of users: > 300 million users

Demographics: > 80% female; 18-35 years

What users do: Use photos for conversation

Type of content: Scrapbooking, pinning photos

Good for: Lead generation, clothing, art, food businesses

Instagram

Total # of users: > 500 million users

Demographics: 18-29 years

What users do: Use photos for conversation

Type of content: images

Good for: Lead generation, retail, entertainment, food

Choose The Right Social Media Platform: Now It's Your Turn

Ask yourself these four questions as you complete today's tasks.

1. Who are my target customers?

2. What social media platforms do these ideal customers use the most?

3. Which social media platforms will probably give me the best return on my time and effort investment (e.g., prospects, leads, new customers, word-of-mouth referrals, etc.)?

4. How do I plan to use the social media platform for my business?

❏ 1. Align customers, purpose, and platform: 30 minutes

Match your target customer to your business purpose, then shortlist the platforms. Use the cheat-sheet above.

❏ 2. Complete your profile and establish your presence: 30 minutes

Once you've identified your preferred social media platform(s), complete your profile and research the most suitable groups, hashtags, or categories. Today's task is to lay a profitable social media foundation so you can start engaging and building relationships in the coming days.

Daily Standup

Did you complete today's tasks?

❏ Yes

❏ No

If no, what do you need to carry over to work on tomorrow?

_____

What did you learn about your business (or yourself) today that will serve you in the future?

_____

60

Day 24 Getting Your Business in Front of 600 Million People

The Platform Every B2B Marketer and Professional Should Use

Without a doubt, LinkedIn is the number-one social media platform for business-to-business (B2B) engagement and professional networking. With 30 million businesses listed and 80 percent of all B2B leads coming from LinkedIn, and with tech skills like cloud computing, software development, and blockchain as most in-demand skills, it's a no-brainer that you use LinkedIn to grow your tech startup.[31]

But maybe I'm a little biased. Back in 2016, LinkedIn was my go-to platform for building my professional credibility and acquiring my first consulting clients. Before I quit my day job, I had fewer than five hundred LinkedIn connections. To be exact, 385. Mostly colleagues. But within a few months of getting active on LinkedIn, I grew my reach to 6,000-plus connections. Those initial clients came from that expanded network. I doubt I would have found them much less have a successful consulting business today if it were not for LinkedIn.

In *The 60-Minute Startup*, I profiled digital marketing agency VisualFizz and its co-founder Dan Salganik, who relies on LinkedIn to grow his business. Dan is a true Agile Entrepreneur—he started his own company and in fewer than thirty days landed his first paying customers. All because of LinkedIn.

Leveraging LinkedIn the Right Way

LinkedIn works if you're looking to reach out to prospective business clients, connect with other professionals, strengthen your professional credibility, switch careers, or learn new skills. Based on my experience and that of other "LinkedInpreneurs" I've interviewed, the right way to find customers on LinkedIn is to:

---

[31] "Eye-Opening LinkedIn Statistics for 2020." Eye-Opening LinkedIn Statistics for 2020. Accessed April 7, 2020. www.99firms.com/blog/linkedin-statistics.

1. **Revisit your ideal customers and their pain points.** On day four, you made a list of your ideal customers and their pain points that your product or service solves. Revisit that list and update as needed. This will be very valuable as you use LinkedIn tools like Sales Navigator to identify your prospects by information (i.e., keywords) on their profile.

2. **Update your own LinkedIn profile:** Make your profile look clean, extensive, professional, and typo-free. Include information about the problems your business can solve, who your ideal customers are, and any testimonials or references you already have.

3. **Produce valuable content:** During the first month of my consulting business, this was my go-to strategy. I wanted to reach as many people as possible with provocative, thoughtful articles. For every comment or share, I engaged with the commenter or shared one of their own articles. I joined a few LinkedIn groups related to my industry and repeated my simple engagement strategy. I sent connection requests to each and every person who liked, commented on, or shared my articles and posts. As a quick aside, my LinkedIn posts (which are about the length of a tweet) had a larger reach than my articles (longer, blog-length content). List articles and how-to articles with a stock image had the best reach of any articles.

4. **Build relationships:** I continued to interact with LinkedIn users who engaged with my published

content. Via direct message, I asked a few questions like:

  a. What are you interested in?

  b. What is important to you?

  c. What problems do you face?

5. **Move the conversation offline**: Once you've built trust and have a chat going, take the relationship off-platform. Suggest a phone call, video call, or an in-person meeting if the prospect is local. More often than not, connections that become offline relationships blossom into paying clients.

Leverage the B2B Sales Engine: Now It's Your Turn

Check off the box beside each task as you complete it.

❏ **1. Revisit your ideal customer and their pain points: 15 minutes**

While revisiting your ideal customer profile, note what job titles they may have, which geographies they may work in, and what type of companies they may work for. You can input this information into LinkedIn's Sales Navigator tool to identify prospects by name. Visit their profiles and start engaging with their content right away.

❏ **2. Join relevant LinkedIn groups: 15 minutes**

Search for groups with keywords relevant for your business. For example, if you're in digital marketing or coaching, join active, engaging digital marketing groups. Just make sure that you finish your profile first so you have a good chance of getting accepted.

❑ 3. Prepare a content calendar and outlines: 30 minutes

This task is to prepare the content you'll publish. Draft a few basic outlines for your first two or three articles and posts. As time allows over the next few days, write and publish them on LinkedIn. Reach out to anyone and everyone who engages with your content. Connect. Ask questions. Get a conversation going. And before you know it, your new connection will wonder how they can hire you.

Daily Standup

Did you complete today's tasks?

❑ Yes

❑ No

If no, what do you need to carry over to work on tomorrow?

_____

What did you learn about your business (or yourself) today that will serve you in the future?

_____

60

Day 25 Borrowing Other People's Customers

How One Tech Ghostwriter Gets Clients Spooky-Fast

In *The 60-Minute Startup*, technical ghostwriter Joshua Lisec set the example for how to find paying customers for a technical service when your own audience is small. Joshua helps authors, consultants, executives, and founders in the tech industry manifest the mission behind their voice. As the world's only award-winning, celebrity-recommended, #1 international bestselling Certified Professional Ghostwriter and founder of www.EntrepreneursWordsmith.com, Joshua "translates" clients' technical expertise into approachable, intuitive, and persuasive articles, books, and speeches.

If you read the first installment in this series, you might remember how Joshua grew his business from $1.67/hour in 2011 to a streak of multiple six-figure revenue years. While other writers apply for low-paying jobs online or mail portfolios to one prospect at a time, Joshua leverages the Other People's Audiences (OPA) Strategy.

Since you're just getting started building your tech side hustle, you probably do not yet have an email list, a customer database, or other "owned" audience to which you can offer your service or product. If that is the case, pull a "Joshua"! Teach other business owners' audiences about your industry as a free value-add and watch as their customers become yours, too.

Borrowing Customers

To execute the OPA Strategy, Joshua connected with accounting firms in his area that serve technology businesses. He asked if their clients experienced a specific problem—content that didn't move their markets. If they said yes, he made an offer. He would teach these accountants' clients something new to grow their business, make the accounting firm look good, and buy everyone lunch. It was a no-brainer deal.

Joshua's workshops were a hit. Accountants won because his presentation got their clients back through their doors, which led to repeat business. Their clients won because Joshua's free presentation helped them fix their flawed messaging. And Joshua won because he positioned himself as the go-to person for technology communications. The first two times Joshua used the OPA Strategy, he spent $250 on lunches total but closed $9,400 in technical ghostwriting work.

The reason Joshua generated business from these lunches was his effective closing. Instead of closing his presentations on a pitch, Joshua said, "If I could demonstrate a process to you where you're able to achieve the results you want, are you open to chatting more? If so, come up, and we'll put a time on our calendars."

A great idea, isn't it? Here's how you can put the OPA Strategy to work for your tech side hustle to find B2B clients.

The OPA Strategy: Now It's Your Turn

Check off the box beside each task as you complete it.

❏ 1. Create your presentation title: 5 minutes

What problem does your product or service solve? What goals do you help customers achieve? You know these like the back of your hand by now. Use a headline to give your presentation an exciting title.

- Could _____ Be Keeping You from _____?

  - ○ *Ex: Could Your Technical Support System Be Keeping You from Retaining Customers?*

- Everything _____ Need to Know about _____

  - ○ *Ex: Everything Executives Need to Know about Building an App for Their Business*

- Learn How to _____ over Free Lunch

  ○ *Ex: Learn How to Leverage Technology to Grow Your Business over Free Lunch*

- _____ Secrets to _____

  ○ *5 Secrets to Get More Done in Your Business Using Low-Cost Cutting-Edge Tech*

- How to _____ (Without _____)

  ○ *How to Increase Your Workforce Productivity (Without Becoming a Terrible Boss)*

❑ **2. List your key presentation points: 5 minutes**

How are you going to deliver what your presentation promises? Take a few minutes now to write down five to ten key steps, tactics, or tips that will allow your audience to digest your advice while digesting lunch.

_____

_____

_____

_____

_____

_____

_____

❏ 3. Design your presentation slideshow: 10 minutes

You don't need a fancy slideshow or beautiful handouts designed by a graphics professional. All you need is a free PowerPoint, Google Slides, or Slideshare template. Put your presentation title on the cover slide. Give each key presentation point its own slide with one accompanying photo. I recommend going to free stock photo websites like Unsplash, Pixabay, and Wikimedia Commons. Your presentation is ready in minutes.

❏ 4. List your potential OPA hosts: 5 minutes

Who do you know that already serves your future customers? Write down three to five potential OPA hosts in the space provided below:

_____

_____

_____

_____

_____

_____

❑ 5. Write your OPA host email pitch: 5 minutes

In Joshua's outreach to accounting firms, he included two messages in a single email. One, a pitch for the OPA opportunity. Two, a pitch for the presentation itself. Below is a template for your OPA pitch based on the exact email Joshua used to book workshops and score big clients.

Hi _____,

Congratulations on your recent successes at _____!

_____ recommended that I reach out to you. He and I have been talking recently about a lunch and learn workshop centered around _____. Have your clients told you about struggles with _____? I thought _____ might be something your clients would like to hear about over a free meal.

Do you agree that a lunch and learn on _____ would be of value to your clients? They'll learn something new about _____, and they'll be grateful because you looked out for their best interest.

If this sounds like a yes or maybe, take a look at this draft email inviting people to the free workshop:

****

Hello (*recipient*),

I'm reaching out to you because we are hosting a lunch and learn on (*topic*). It's called (*your presentation title*).

(*your title and name*) let us know about a sobering statistic:

(*a statistic that proves the need for your product or service*)

Because of (*comment on the statistic and give your opinion of how it affects your potential customers*)

While you enjoy a free lunch, (*your name*) will be showing you how to:

● (*your key presentation points*)

If you want to stop (*a problem your product or service solves*) and (*a benefit of working with you*), please call our office at _____ to reserve your spot.

****

If this looks like a go, let me know the best date in (*month*) or (*month*) to get this lunch and learn on your schedule.

❏ 6. Order boxed lunches: 10 minutes

NOTE: You will complete steps six and seven on the day of the presentation itself. Obviously I'm not expecting you to complete all seven steps today! It may take a few days or a week to hear back from your potential OPA hosts and schedule your presentation. Do everything you can now and finish the final steps as you book your lunch and learns.

Give your presentation's attendees a respectable lunch. Fast casual fare is fine, fast food is not. When Joshua used the OPA strategy, the CPA firms' executive assistant took each guest's lunch order when they called to reserve their spot.

❏ 7. Deliver your presentation and get customers: 20 minutes

Dress business casual. Test your projector and presentation as soon as you arrive. Greet everyone with a firm handshake, give each guest their lunch, and ask them to hold any questions until the end. To get the ball rolling, ask everyone what they hope to take away from the presentation. You might keep those expectations in the back of your head so you fulfill them in your talk.

Deliver your presentation at a slower speed than feels normal. Slowing down helps you keep uhs and ums to a minimum. Smile at and talk to your audience, not your slides. Keep your talk to fifteen minutes so people don't get bored or leave to get back to work.

Daily Standup

Did you complete today's tasks*?

❏ Yes

❏ No

*Bookmark this page and come back to check off the yes box when you've completed steps six and seven.

If no, what do you need to carry over to work on tomorrow?

_____

What did you learn about your business (or yourself) today that will serve you in the future?

_____

60

Day 26 A Free Offer to Get Paying Customers

Free + Premium = Freemium

Way back in 2003, Niklas Zennstrom and Janus Friis started a small company with a funny name: Skype. It used an exciting new technology called Voice over IP (VoIP) to call other people over the internet. The software was free to use, which led to its immediate adoption by hundreds of millions of internet users. Skype got so popular that eBay bought the company for $2.6 billion in 2005.

Former food stamp welfare recipient Jan Koum and fellow Yahoo ex-employee Brian Acton applied for open jobs at a young company called Facebook but did not make it to the final round of interviews. To pay the bills, they brainstormed business ideas they could start quickly and with very little upfront capital. Together with a third friend, Alex Fishman, and programmed Igor Solemennkov, whom they found on www.RentACoder.com, they built WhatsApp. They made the text messaging app available as a free download. By 2014, 200 million-plus people were using WhatsApp. That same year, Facebook acquired the company for $19 billion.[32]

On day four, you learned that Nathan Miller gave away his property management software RentecDirect for free for months before charging for an upgraded version. William and John from day eight use the freemium strategy for WeWorked. The barebones software was free to use, but people need to pay for advanced features and customizations. There's a lesson in these stories, and today, we're going to learn it.

When Can Free (Or Freemium) Work?

Freemium is a business model where a company offers a version of its subscription-based service to all customers for free. The free version is usually a basic version of the bells-and-whistles paid service. The idea is that once

---

[32] Olson, Parmy. "Exclusive: The Rags-To-Riches Tale Of How Jan Koum Built WhatsApp Into Facebook's New $19 Billion Baby." *Forbes Magazine*, April 23, 2014. www.forbes.com/sites/parmyolson/2014/02/19/exclusive-inside-story-how-jan-koum-built-whatsapp-into-facebooks-new-19-billion-baby.

customers fall in love with the free version, a certain percentage of free users will decide to purchase a subscription for the robust version. At the very least, they'll spread the word about the free version to others. On and on the cycle goes.

Since your business is new to the market and probably doesn't yet have a track record, offering a freemium version is a quick way to get trial subscribers. It's a low barrier to entry ("If I don't like, at least I didn't waste any money.") A freemium offer also makes a product shareable, which enables easy referrals and therefore more signups. The bigger your pool of free users, the more likely several will convert into premium customers.

Ask yourself these questions to determine if the freemium model is right for your business:

1. Can I offer a low cost-to-serve service to support the free users?

2. Is there a clear path for free users to become paying users?

3. Does my service have a *huge* potential market?

4. Does my service's value increase the longer people use it so users can't "switch" later on?

5. Does my service have viral adoption potential (i.e., each user is rewarded for encouraging other users to join)?

6. Can I monetize free subscribers through advertisements?

Here are five ways you can leverage the freemium business model to acquire customers:

1. Limited Time Free Trial: # days free, then user pays

2. Limited Features: Basic version free with a sophisticated paid version

3. Limited Seating: # people can use for free, but any number over that must pay

4. Limited Customer Type: Small companies can use for free, but big companies must pay

5. Limited Storage/Capacity: Certain storage space is free, but more than that requires a paid subscription

Implement the Freemium Model: Now It's Your Turn

Check off the box beside each task as you complete it.

❏ **1. Assess if your product or service is right for freemium: 15 minutes**

Using the self-assessment questions as your guide, decide if freemium is the right model for your business.

❏ **2. Implement the appropriate freemium model: 45 minutes**

Of the five freemium models, choose the one most appropriate for your business.

Daily Standup

Did you complete today's tasks?

❏ Yes

❏ No

If no, what do you need to carry over to work on tomorrow?

_____

What did you learn about your business (or yourself) today that will serve you in the future?

_____

Day 27 Focus on Tech-Only Online Communities

The What and Why of Tech-Only Online Communities

On day nineteen, I wrote about finding your internet tribe and specifically about using Quora and Reddit to find your customers. Quora and Reddit can be good hangouts for all types of tech startups, including services like consulting and coaching as well as products. But if you have a tech product to market, there are some excellent online communities just for tech products, and especially SaaS products.

The numero uno in this area is a community called Product Hunt. You may have heard of Slack, a company with a $15B market cap, which has a very popular chat room app. It's designed to replace email as a primary method of communication and sharing. Slack became popular via Product Hunt. Other companies that became popular via Product Hunt include Robinhood (a $0 commission stock brokerage) and Digit (an SMS bot that monitors your bank account and saves you money).

If you're a product enthusiast, you've probably already heard of Product Hunt. It shares and discusses the best and latest apps, websites, and technology. Product Hunt features separate tabs for games, books, and podcasts. Invites are only granted to the most active members, so if you go there simply to promote your own product without participating in the community, you can't. If you're looking for feedback and traction for your startup, then Product Hunt is the right place. Other features of Product Hunt you might find useful include Meetups, Product Hunt Radio, and Live Chats with entrepreneurs.

Launching a new tech product or startup isn't exactly a walk in the park. Every new product must overcome a litany of challenges, with arguably the biggest one being finding and connecting with a relevant audience. Every new product needs to find its market to succeed. There are a bunch of great tactics and strategies to help you do this, but one of the best ways to promote your launch, connect with an audience, and find your evangelists is by launching on Product Hunt.

Another popular community for tech startups is Hacker News. Hacker News is a place to discuss startups and technology. It's more focused and geared towards startups that are well invested in and present in Silicon Valley. Making it to the front page of Hacker News is considered a big deal considering their strict community guidelines and focus on quality posts. You can quiz fellow community members about anything startup-related on the Ask HN tab and receive feedback on the Show HN tab. Other tech platforms include GrowthHackers and Online Geniuses.

The "How To" for Online Communities

Now that you have a good understanding of these tech-focused communities, let's go over the best known methods to help get you your first customers. We'll use 'Product Hunt' as an example.

1. Prepare your 'hunter kit."

    a. Enter the name of the product (max 60 characters)

    b. Include the URL you want to promote. If you have app URLs as well, put them here, too.

    c. Enter your tagline (max 60 characters)

    d. Include the platforms your product runs on

    e. Upload five to ten images showcasing different parts of your product. If you also have a YouTube video, that's perfect for here, too!

    f. Tell them what freebies you're preparing to give to the Product Hunt community, if any

2. Find a "hunter." This is someone who has a decent following and who can actually launch the product on Product Hunt. Finding a good hunter and building a relationship with them might take some time, but it'll be time well spent.

3. Announce your listing to your followers so they can upvote your product. Don't ask them directly to upvote it (which is a no-no). Just inform them about the listing.

4. Comment and reply to comments to keep yourself engaged with your product followers.

Online Communities: Now It's Your Turn

Check off the box beside each task as you complete it.

❏ **1. Complete your profile and a product profile in an online community: 30 minutes**

This is the most important piece you need to work on today. If your profile or your product profile is not complete, all your efforts will be wasted!

❏ **2. Reach out to influencers and followers who could support your launch: 30 minutes**

Research other influencers who may be interested in your product or service by looking into their past posts. Then put together a list of followers who can support your launch.

Daily Standup

Did you complete today's tasks?

❏ Yes

❏ No

If no, what do you need to carry over to work on tomorrow?

_____

What did you learn about your business (or yourself) today that will serve you in the future?

_____

6🕐

Day 28 The Trade Show Gold Mine

―――――――――――――

Mining Trade Shows and Conferences for Customers

When you think of trade shows and conferences, you probably imagine thousands of people, miles of booths, and standing room-only workshops. But if you're strategic about in-person industry events, they can offer a treasure of prospects with whom you can make a persuasive first impression.

That's exactly what Vivek Kumar did. Vivek is the founder and CEO of Qlicket, which helps employers reduce personnel turnover and gets employee buy-in to address issues inside the company. Early on in his business, Vivek identified relevant conferences where his target market would be, and he got the attendee list from event administrators. He then emailed all attendees requesting a lunch meeting to resolve their employee turnover problems. Anyone who responded with a yes Vivek emailed an explainer video about his solution. This strategy gave Vivek his first business breakthrough—massive lead generation and big revenue.

Cory Minton from day one travelled extensively for his work as a technology evangelist. That's how he met data influencers whom he would later interview for his Big Data Beard podcast, his side business. Data influencers also liked the fact that they had an opportunity to speak to Cory's fast-growing audience. Some of these influencers later became paying customers via podcast sponsorship. It all started with a handshake.

The Step-by-Step Approach to Find Paying Customers in Person

By now, you have a good idea of your ideal customers, their pain points, and the ways your product or service uniquely solves them. The next step on your agile entrepreneur journey is to choose conferences your ideal customers visit and connect with them using both online and offline methods. Here are the steps.

1. Research and select the right shows to attend. There are hundreds of shows that are happening around the world during any given month. Here are a few

websites I found useful. Select based on your budget and proximity:

    a. www.Blog.Bizzabo.com/Technology-Events

    b. www.TechConferences.co

    c. www.TechMeme.com/events

2. Get the attendee list the traditional way. This is the toughest part of the exercise. Some organizers share the visitor information either on their website or the event app. You may have to manually collect this information. For sure you'll have information on exhibitors and speakers, which is a good start.

3. Get the attendee list the social media way. Post on your favorite social media platforms that you are planning to attend the conference and would like to meet other attendees. They may not be your ideal customers, but you'll have a start.

4. Connect with other attendees. Email attendees and offer to meet for coffee, lunch, or dinner during the event.

5. Connect with other potential customers by asking your original list to help you connect.

6. Follow up afterwards.

Set up Your Conference Schedule and Start Emailing

Check off the box beside each task as you complete it.

❏ 1. Create a list of important conferences and trade shows in your niche: 15 minutes

Every niche has important conferences and trade shows. Do a quick Google search and make a list of those that are most important to your customer base. Look for well-known industry leaders and celebrities who are featured speakers. Chances are, that event will attract people with money—people who might just be perfect customers for your business.

❏ 2. Select two or three that fit your schedule and budget: 10 minutes

Choose two or three conferences or trade shows that are happening soon and near you. Make sure the ticket prices are within your budget. Buy the tickets, put the dates in your calendar, and put a smile on your face. You're about to find customers!

❏ 3. Get the conference attendee list: 10 minutes

Go to the conference website, find out who is organizing the thing, and reach out to them via email or social media. Ask for a list of attendees. Some organizers give this list away for free, while others ask you to pay for it.

❏ **4. Email attendees and ask to meet in person: 25 minutes**

Send out individual emails asking to meet attendees in person. Don't BCC the entire attendee list. Individual emails, personalized messages—these work best. Include a brief note about who your business helps and how in your email so recipients can self-identify that they're a fit for your product or service. It would not surprise me if you make a sale before you ever meet anyone in person!

Daily Standup

Did you complete today's tasks?

❏ Yes

❏ No

If no, what do you need to carry over to work on tomorrow?

_____

60

Day 29 Negotiation and Objection-Handling

Start with Objections, End with Negotiations

Negotiations happen when you lay out your terms to work with you, they differ from the client's, but you both want to find a way to work together. Objection-handling, on the other hand, addresses your prospect's concerns about your product or service to get them to understand the technical aspects and achieve their buy-in. Both of these come after you send a proposal either via email or face-to-face during the sales call. Some prospective clients use objections and negotiation tricks as tactics to get a better deal; others simply have questions or concerns they need cleared up before they can give a wholehearted yes.

The fundamental rule of negotiating is always the same whether you're talking to partners, customers, or investors. It's never about beating someone. It's about creating the best possible deal for all parties involved. The key is to determine the value of what you're offering, find a consensus, and make both sides of the table happy—a win-win.

People think negotiation is all about price. But a discount is usually not what makes customers buy. It's the value. What does your customer need? What tangible difference will your product or service bring? Is its value worth the price?

Whether you're talking price or some other customer concern, it's better to negotiate only once you've established a fit for both sides. You don't want to "win" them over only to have them buy the product for all the wrong reasons . . . and soon demand a refund.

Zig Ziglar was one of the greatest salesmen and negotiators of all time. Here are his three negotiation principles:

1. Have an absolute and total belief that what you're selling is worth more than the price you ask for it.

2. Mentally prepare yourself.

3. Use emotion and logic in your presentation.

I interviewed and profiled Sales Acceleration Group founder Kristie Jones for *The 60-Minute Startup*. Kristie helps small and mid-sized businesses with sales strategies and sales development. To prepare for inevitable negotiations with potential clients, Kristie always sends a proposal with clear expectations on her services' value before ever discussing price. Once the prospect understands the value and they agree they're a good fit, any negotiations simply focus on logistics and next steps rather than sticky issues like price.

Clarity Is Key

While **negotiating,** keep in mind these four important points:

1. Always find the right decision-maker. Can the prospect you are talking to make decisions? Don't waste your time or money if you can't find the decision maker.

2. Thorough preparation of your prospect. In addition to being up to speed on your offer, make sure that you know thoroughly about your customer, their pain points, their business situation, and why they may be interested in your offer.

3. Be clear on your strategy. What are your goals? What is the price point that you can't afford to sell at? Can they give you name recognition even if you sell for a lower price? What can the customer get you in the long term?

4. Do not oversell. Always talk about the value customers will get and the benefits of your offer.

There are three primary **types of objections** you'll face in your negotiations:

1. The prospect lacks interest. This is not an objection in its true sense, but you'll notice that the prospect you are talking to does not seem to have much interest in your product or service or the negotiations. More than likely, the reason for this is that the qualification process of showing value in the services and identifying the right person is not done well. It is better to find the right decision-maker or find a creative way to demonstrate value.

2. The timing isn't right. Find out about your prospect's long-term and short-term goals. Is there a subset of your service or product that can meet their short-term goals? Do not force the sale but rather nurture the relationship.

3. There are budget constraints. This means the prospect sees the value of your products or services less than the cost. Try pointing out the return on investment (ROI) or offer some flexibility like a free trial period. Or throw in an additional bonus period for your subscription.

Prepare Your Negotiation Strategies and Objection Checklist: Now It's Your Turn

By this time, you would have engaged with multiple prospects and may already have a few customers in the pipeline. Check off the box beside each task as you complete it.

❑ 1. Create a list of negotiation strategies: 30 minutes

Based on your experiences so far, create a list of negotiation strategies you have used, the outcome of each of those strategies with specific examples. If you do not know where to start, head on over to www.The60MinuteStartup.com for a template with some pre-filled strategies that you can take advantage of.

❑ 2. Prepare list of objections and your answers: 30 minutes

Similarly, prepare a list of objections you have heard in your engagements so far and how you addressed them. If you need help, download a pre-filled template with common objections to tech products and services from www.The60MinuteStartup.com.

Daily Standup

Did you complete today's tasks?

❑ Yes

❑ No

If no, what do you need to carry over to work on tomorrow?

_____

What did you learn about your business (or yourself) today that will serve you in the future?

_____

60

Day 30 Grow Your Business
With Referrals

Getting Customers to Market Your Business for You

More than eight in ten people say they trust product recommendations from family and friends.[33] From an entrepreneur's perspective, receiving these recommendations in the form of referrals is one of the most effective ways to grow a business.

---

[33] "Global Trust in Advertising – 2015." Nielsen, September 28, 2015. www.nielsen.com/us/en/insights/report/2015/global-trust-in-advertising-2015.

Customer referrals make up roughly 20 percent of Brian Meert's business at his advertising agency AdvertiseMint.com. When Brian's first customer—his employer—hit tough times and could no longer afford his services, they referred other companies to Brian so he could replace that revenue several times over. Cory Minton grew his Big Data Beard podcast audience through referrals from colleagues to other people.

So, how do you get people to send you customers? To do that, we must understand the psychology of word-of-mouth marketing. People give referrals the same way they buy things. When someone decides a product or service is the right fit and the right price—logic—they must also believe the purchase will make them look good, impress people, or boost their quality of life—emotion—before they buy. Usually this happens in reverse order. Buyers get hooked on the emotion of a purchasing decision and back it up with logic.

To make your business worthy of referral, tap the logic and emotion formula. That is, people have to *believe* you can help, that you offer a great price, and that you will show up and deliver as promised. That's the logic. But they must also feel good about helping you, trusting that their friend will be treated well. That's the emotion.

This phenomenon explains why cash-for-referrals is a poor motivator. Maybe you tell family and friends that you'll reward them for any referrals they send you, but the prospects they send your way will probably not be worth your time. Far better to make your business more likable, more referable, and more energetic before offering any referral incentives. Today is all about doing just that.

The How-To for Setting up a Referral System

1. Look for opportunities for a positive response

2. Provide a template

3. Act on positive feedback

4. Distribute your content and resources

5. Create different avenues for advocacy

6. Add a customer loyalty program

7. Exceed expectations

8. Adopt a customer referral program

9. Refer other companies

10. Offer incentives

Establish A Referral System: Now It's Your Turn

Check off the box beside each task as you complete it.

❏ **1. Prepare a list of people who might refer your business: 15 minutes**

Prepare a list of people (family, friends, colleagues, customers, etc.) who believe in you and your business. They must also feel comfortable referring their network to your business.

❏ **2. Write a referral script to make it easy for contacts to refer: 30 minutes**

Make it easy for people to send you referrals. Write a referral script that your potential word-of-mouth marketers can use to let people know about your business. This script can take the form of a short social media post, an email, or just a few sentences that people can bring out naturally during conversation.

❏ **3. Prepare a list of people you can refer business to: 15 minutes**

To encourage others to refer to your business, practice what you preach. Send them referrals! This generates goodwill, which goes a long way to boosting referrals to your business.

Daily Standup

Did you complete today's tasks?

❏ Yes

❏ No

If no, what do you need to carry over to work on tomorrow?

_____

What did you learn about your business (or yourself) today that will serve you in the future?

_____

The Finish Line (Or Is It?)

Wow, you've made it! You have just crossed the thirty-day marathon finish line after sprinting daily for sixty minutes. That is an accomplishment worth celebrating!

If you have just been reading the book first to understand the concepts before actually implementing, congratulations are still in order for going over all the key concepts of the agile entrepreneur's way to start a business in thirty days or fewer and get paying customers.

If you have implemented the strategies mentioned in the book for the past month or so, you are most likely looking at one of the following three scenarios.

- **Scenario One**: You've launched your business and now have at least one paying customer thanks to the strategies covered in the last two weeks.

- **Scenario Two**: You launched your business and tried your hand at a few marketing strategies but are still waiting on your first deal to close.

- **Scenario Three**: You were unable to launch your business because of some unexpected roadblocks.

For all three scenarios, the journey is far from over. There are multiple turns you can (and will) take to continue on your current path. If scenario one is your story, you have made a great accomplishment. You've overcome the most significant hurdle any entrepreneur will face—getting that first paying customer. Now your task is to go after the next set of customers. The same marketing methods you've used already will help you add to your revenue, then multiply it going forward. The most important strategy is customer referral generation from existing customers. Ask for testimonials and referrals, and don't be shy about it.

For scenario two, your next key task is to learn from what failed. Which day's activities didn't pan out? Which strategy went wrong? Why? Many times, the strategy itself may not be the problem, it simply needs a little more time to work. Could that be the case in your situation? Or do you need to focus on other strategies mentioned in the book that you didn't get the chance to implement? Or both? In either case, reflect on your journey with a positive attitude. You now have a business that you can call your own. Use that reality to your advantage to promote your business in ways that you *know* work for other entrepreneurs just like you.

If scenario three best describes your situation, then you my friend have the easiest problem to fix. If you were unable to launch your business, more than likely the cause was inadequate resources. Not enough time, no money, not the right skills. So let's first figure out what is the underlying cause of why you did launch your business. Can you fix that issue? If you didn't have enough time, do you think you can find time in your schedule to go back and do the activities you skipped? If money was the problem, are you in a position to shore up finances without going broke? If it's a missing skill, can you outsource or otherwise get help from family or friends to fill that gap? Perhaps someone you know who has the skills your business needs is thinking about starting a business. In that case, the simple solution that changes everything for you might be a partnership.

The entrepreneurial journey is exhilarating and exhausting. Many veteran entrepreneurs will admit that the happiest and most depressing times of their lives have been during their business journey. Sometimes both at the same time. My effort with this book is to give this journey more balance. I want you to focus only on activities that lead to paying customers. After all, that is the only goal of a business, as Peter Drucker said. Rid yourself of all other noises. Keep asking yourself what innovation and what marketing you could do to create your next paying customer.

And if you need help, I am just one website away at www.The60MinuteStartup.com. If you have any questions, feel free to reach out to me at Contact@The60MinuteStartup.com. I promise to respond to you as soon as I can (you can connect with me on LinkedIn as well).

Thank you for reading *The 60-Minute Tech Startup*, and congratulations on reaching this far. Your exhilarating journey has only begun!

What's Next?

www.The60MinuteStartup.com

Start your business today the agile way! As a valued reader, you get all the templates, scripts, and tools you need to build your 60-minute startup, all at no cost to you.

Just visit this special web page for your free content upgrades so you can set up your business, build a website and other online assets, attract paying customers, and more in the shortest time possible.

Want support from other entrepreneurs on this thirty-day journey? Join the free private group to ask questions, bounce around ideas, and possibly even find your first paying customers. Go to the link below.

www.The60MinuteStartup.com

About the Author

Ramesh Dontha is a serial entrepreneur, host of The Agile Entrepreneur Podcast, and author of The 60-Minute Startup Series, which includes *The 60-Minute Startup* and *The 60-Minute Tech Startup: How to Start a Tech Company as a Side Hustle in One Hour a Day and Get Customers in Thirty Days (or Less)*. As a manager and consultant for Fortune 100 companies, Ramesh used the Agile Methodology to make technology systems more efficient. He then applied Agile principles to entrepreneurship, starting, growing, and selling multiple technology businesses. Now Ramesh teaches side hustlers how to get paying customers for their tech businesses. Start your tech business as a side hustle in sixty minutes a day at www.The60MinuteStartup.com.

Printed in Great Britain
by Amazon

87683231R00128